A TEXTBOOK ON
ENVIRONMENTAL STUDIES

A TEXTBOOK ON
ENVIRONMENTAL STUDIES

By

DR. RAJAN MISHRA

M.A. Sociology (A.C.), M.S.W., N.E.T., Ph.D.

Director, Dr. Bhimrao Ambedkar Polytechnic
Agra (U.P.)

UNIVERSITY SCIENCE PRESS

(An Imprint of Laxmi Publications Pvt. Ltd.)

**BANGALORE • CHENNAI • COCHIN • GUWAHATI • HYDERABAD
JALANDHAR • KOLKATA • LUCKNOW • MUMBAI • PATNA
RANCHI • NEW DELHI**

Published by :

UNIVERSITY SCIENCE PRESS

(An Imprint of Laxmi Publications Pvt. Ltd.)

113, Golden House, Daryaganj,
New Delhi-110002

Phone : 011-43 53 25 00
Fax : 011-43 53 25 28

www.laxmipublications.com
info@laxmipublications.com

First Edition : 2009

OFFICES

✆ **Bangalore**	080-26 75 69 30	✆ **Chennai**	044-24 34 47 26
✆ **Cochin**	0484-237 70 04, 405 13 03	✆ **Guwahati**	0361-251 36 69, 251 38 81
✆ **Hyderabad**	040-24 65 23 33	✆ **Jalandhar**	0181-222 12 72
✆ **Kolkata**	033-22 27 43 84	✆ **Lucknow**	0522-220 99 16
✆ **Mumbai**	022-24 91 54 15, 24 92 78 69	✆ **Patna**	0612-230 00 97
✆ **Ranchi**	0651-221 47, 64, 220 44 64		

UES–9311–120–ATB ON ENVIRON STUDIES **C—16520/08/09**

Typeset at : ABRO Enterprises, Delhi. *Printed at:* Repro Knowledgecast Limited, Thane

Dedicated in Loving Memory of My Father
Late Shri S.K. Mishra
(Ex. Prof. Agronomy)
&
My Mother
Late Smt. Savitri Mishra

CONTENTS

PREFACE

The book, 'Environmental Studies' has been written in a simple language. It focuses on Multidisciplinary Nature, Environment, Natural Resources, Ecosystem, Biodiversity, Environmental Pollution, Social Issues and Human Population and the Environment.

All the chapters have been supplied with neat and clean diagrams which help the students to understand the text. Glossary and Appendix have been given at the end of this book which has also useful for the students to understand difficult words.

In presenting this book to the students, I hope that this book will be quite helpful for acquiring the skills of various environmental factors. The subject Environmental Studies has now also been introduced in Indian Universities. Every attempt has been made to familiarise the students with the latest developments taking place in the theory as well as field work of "Environmental Studies."

The charts and tables has been extended to a greater extent to make the subject material more readable and to help the students in grasping it clearly. Case studies and suitable examples have been cited to ensure clarity of the subject.

I owe a deep sense of gratitude to my friends, colleagues and students for providing me enough opportunity for interacting with them.

I hope that this book will be appreciated by my learned colleagues and students. Any constructive suggestions regarding the improvement of this book will be acknowledged.

—Author

ACKNOWLEDGEMENT

My special thanks to Dr. Nistha Dubey, Lecturer, Sachdeva Institute and Technology, Agra, Mr. Sanjeev Singh Thakur, Anand Engineering College, Agra, Mr. Rishi Kapoor, Educational Consultant, Agra, Naveen Prakash, Sr. Lecturer, College of Engineering and Technology, Moradabad, Er. (Prof.) Sharadha Saxena, Government Polytechnic, Lucknow, Er. Rahul Singh, Lecturer—Computer Science, F.E.T.R.B.S., Dr. Vipin Agarwal, Reader, R.B.S., Bichpuri, Agra, Dr. Aneesh Satrangi, Lecturer—Management, M.D. College, Agra, Mr. Satendra Yadav, Lecturer—Management, G.L.A. Mathura, Ms. Richa Kurukshetra, T & P Officer, S.I.T., Mathura.

—Author

Unit I ENVIRONMENT

The term 'Environment' means surroundings, in which the organisms live. It is the sum total of all living (biotic) and non-living (biotic) factors that surround and potentially influence the organism. Some components of the environment serve as resource (*e.g.,* water, soil etc.) while other act as regulatory factor (*e.g.,* light, temperature etc.). Different components of the environment are interdependent and interlinked. Environment creates favorable conditions for the existence and development of living organisms. In context of human being, environment can be defined in number of ways:

(*i*) *"Environment is the representative of physical components of earth wherein man is the important factor influencing the environment".*

(*ii*) *"Environment refers to the sum total of conditions which surround man at a given point in space and time."*

(*iii*) *"Environment is the sum of all social, economical, biological, physical or chemicals factor which constitute the surroundings of man, who is both creator and molder of his environment."*

The environment can be studied at both large and small scales. This is reflected in regional and global climatic patterns, as well as the local climatic conditions called the microclimate. The term 'spatial' refers to the physical space occupied by an organism.

Most organisms interact with their environment at several spatial and time scales *e.g.,* a single bacterium in soil interact with air and water within a fraction of a cubic centimeter space, while a tree interact with a large volume of air, water and soil at a large spatial scale. The environment varies from place to place to variation in climate and topography. The activities of organisms influence the hydrosphere, the lower atmosphere and the near surface part of the lithosphere through exchanges of matter and energy. Organisms have to cope with the external environment over a range of time scale, varying few minutes to days, seasons or over a much longer period of geological time scale. For example, the population of phytoplankton (*i.e.,* microscopic photosynthetic organisms that float on water surface) may undergo change with a few days the change in light conditions in aquatic systems. On the contrary, the variations in lithosphere occur very slowly and take a long period of time.

Necessity of Environmental Education. Environmental education is the process of recognizing value and clarifying concepts in order to develop skills and attitudes necessary to understand and appreciate the interrelates of man, his culture and his biophysical surrounding. It also entails practice in decision making and self-formulation of a code of behaviour about issues concerning environment quality.

Environmental Studies. Now most of the countries, including India have realized the need of environmental education. The main objective is that the individual and social groups should acquire awareness and knowledge, develop skills and abilities and participate in solving real life environmental problems. The prospective should be integrated, inter-disciplinary and holistic in character.

NEED AND IMPORTANCE OF ENVIRONMENT

Every living species of plants or animals influences its environment and in turn gets influenced by it. The magnitude of such influences is not usually high in these species because of the fact that due to natural checks their population cannot rise beyond certain limits and they can also not modify their own way of life. However, man is an exception. With increasing scientific knowledge, man is able to modify the environment to suit his immediate needs much more than any other organsim. This enables man to improve the quality of his life.

Since the very beginning of human civilization, some thousand years ago, man started interfering with the environment. He devastated forests by cutting trees for the wood and for other household needs. He removed stretches of forests for bringing land under cultivation, only to find his water supply diminishing and his supply of soil eroding away. He killed animals, the gentle ones for food and the fierce once for safety. He had polluted the rivers with chemicals from factories, there by making the water unfit for his needs. All these activities, however, did not affect the environment too seriously up to a fairly long time, because the population was not too high and lifestyle was not so complex. The natural self-purifying and self-cleaning capacity of the environment was underteriorated.

After the scientific and industrial revolution in the recent past, there has been immense impact of man on his environment. Man has failed to relies that any new factor upsets the balance of the ecosystem as a whole. All the devastating effects of man's efforts to control nature have occurred because he has upset the balance relationship of the organisms that make up the environment. Huge industrial installations every year, introduction of faster mode of transport and sprouting up of large crowded cities and fast urbanization are the main outcomes of the modern civilization. These and a large number of many others are contributing to what is called **environmental pollution,** an example of which is also the widespread use of insecticides in USA during 1950s. The immediate effect was a reduction in the population of pests and an increase in the yield of crops. But these insecticides also poisoned and killed birds which feed on insects. As a result, the next generation of pests multiplied even faster in the absence of their natural enemies and damage to crops was much more increasing industrialization is also causing much danger to man's life by polluting the environment.

TYPE OF ENVIRONMENT

Environment can be divided into two categories:

(*i*) *Natural Environment.* The environment that comes into existence without interference of man is called natural environment. It operates through self-regulating mechanism called homeostatic environment mechanism *i.e.*, any change in the natural ecosystem brought about by natural process is counter balanced by changes in other components of the environment.

(*ii*) *Man-made or Anthropogenic Environment.* The environment which has been modified by human activities is called man-made environment. Man is the highest of all creatures on this earth. He is modifying the environment according to his own needs and ways without taking into account its consequence. Increase in the scientific technology which is the products of human brain is now deteriorating the environment.

COMPONENTS OF ENVIRONMENT

The environment can be classified into two components:

1. Abiotic (non-living)
2. Biotic (living)

1. Abiotic (Non-living) Components

The abiotic or non-living components include the **climatic** and **edaphic factors.** The climatic factors include temperature, rain, humidity and snowfall etc. The edaphic factors comprise the soil and substratum. The ability of organisms to utilise, tolerate or combat the various abiotic factors are different and it may limit their distribution, behaviour and relationship with other organisms. Some of the most important aboitic factors are given below:

(*i*) *Background.* The background of the habitat also determines the distribution of animals by enabling them to camouflage against the colour, general texture and patterns *e.g.*, desert animals like the lion and the camel are sand coloured.

(*ii*) *Humidity.* It is the amount of water vapours present in the atmosphere. Humidity regulates the rate at which water evaporates from the body surface of land organisms by transpiration, perspiration and other means. Different plants and animals show various adaptations to withstand dry conditions.

(*iii*) *Light.* Light is essential for photosynthetic organisms for the preparation of food on which the rest of living world depends. Plants show various adaptations for obtaining optimum light. The activities of a large number of organisms are regulated by light *e.g.*, moths, cockroaches and bats are active during the night.

(*iv*) *Mineral Elements.* A large number of minerals are required by organisms for their proper growth. Deficiency or absence of anyone results in abnormal growth which may lead to death. High concentration of minerals are equally harmful and can limit the distribution of organisms.

(*v*) *Temperature.* Living organisms can survive only in a narrow range of temperature which allows their metabolism. The living organisms develop physiological and behavioural adaptations to withstand extremes of temperature. Many animal (such as bird and mammals) migrate to warmer places in winter to avoid extreme cold. Some desert animals live inside burrows to avoid the intense heat of the desert.

(*vi*) *Water.* Water is essential for life. No life can exist without water. The requirement of water varies from organism to organism. The distribution of organisms depends upon the extent of the need and special adaptation for conserving water.

(*vii*) *Wind.* Wind determines the weather condition. It helps in the dispersal of seeds and fruits of many plants.

2. Biotic (Living) Components

The living organisms form the biotic components of the environment. All organisms require energy for their life processes and materials for the formation and maintenance of body structures. Food supplies both energy and materials for the subsistence of life. Green plants produce carbohydrates by photosynthesis and also synthesise proteins and fats. The green plant are, therefore, called **Producers.** The animals that consumes this energy as food are known as **Consumers.** The non-green organisms like the fungi and some bacteria which are incapable of producting their food, live on the dead and decaying plant or animal parts are known as **Decomposers.**

Every living species of plants or animals influences environment and in turn gets influenced by it. The magnitude of such influences is not usually high in these species because of fact that due to natural checks their pollution cannot rise beyond certain limits and they can also not modify their own way of life. However, man is an exception. With increasing scientific knowledge, man is able to modify the environment to suit immediate needs much more than any other organism. This enable man to improve the quality of his life.

Since the very beginning of human civilization, man has started interfering with the environment. He devastated forests by cutting trees for the wood and for other household needs. Man removed stretches of forests for bringing land under cultivation only to find his water supply diminishing and his supply of soil eroding away. He killed animals for food and the fierce ones, for safety. He had polluted the rivers with chemicals from factories thereby making the water unfit for his needs.

In the recent past, after the scientific and industrial development, there has been immense impact of man on his environment. Man has failed to realize that any new factor upsets the balance of ecosystem as a whole. Huge industrialization, introduction of faster mode of transportation and sprouting up of large crowded cities are the main outcomes of modern civilization. These and a large number of many other causes are contributing to what is called environmental pollution.

NEED FOR PUBLIC AWARENESS

With the advancement in technology, explosive increase in human population, scarcity of space and food, deterioration of hygienic conditions, depletion of natural resources and socio-economic problems, one needs extensive and exhaustive study of the environment, particularly in relation to human survival and benefit. The subject is gaining more popularity with its name as environmental biology or environment studies. However, the environmental biology is not a new discipline but simple an extension of ecological approach which stresses upon the study of environment in its totality with special emphasis on the welfare of man.

Environment is not merely the atmosphere and other physical factors, surroundings, but is the complex of all factors which not only affect one organism, one time, but all organisms all the time. In a larger sense, environment constitutes the various physical, mental, social, spiritual, educational, economic and intellectual aspects of whole humanity and when kept healthy and inspiring, it promotes the progress and development of mankind. The understanding of the subject needs an adequate knowledge of geography, climatology, pedology (soil science), microbiology, biochemistry, biophysics, sociology, economics, etc. The so called progress, in reality, is disturbing the environmental balance seriously, leading to breakdown of life support system on the earth. The toughless exploitation is mainly due to the ignorance about environment and ecosystem, lack of concern amongst planners and ecologists about the side-effects of the developmental projects, gap in information and in training personal, absence of multidisciplinary approach to development project and lack of systematic information, data and feedback system. These calls for making proper environmental strategies for maintaining the ecological balance not only by making preservation of nature an integral part of development planning, but also by adopting alternative means of livelihood for those dependent upon nature for survival.

Till sometime before, when knowledge was accountable to some main disciplines *e.g.*, literature, history, geography, science etc. environment was not one of them. But as the claws of industrialization and complications of modern lifestyle began to shadow the world consciousness roused vehemently in respect of environment. Resulted, thereby environment could secure a separate seat in the row of disciplines already studied. Furthermore when it was taken into notice that the environment was also an influential force, responsible for the conclusions during the course of studies in respect of certain other disciplines, it was also added to them. This naturally refers to the scope of environment and also of its being a multi-natured discipline.

Usually give diversion of the venue of humanity. However, this power is not generally considered by the man to be in his sphere of botherance, because the human being has boons of self-thinking and decision-taking power about his activities. This makes the man physically and mentally active on every account of life. Thus, he thinks himself free and not enslaved. But it is not true that he is not bound by the social, political and religious net which he has knitted himself for him to live in. It is therefore, the man does not consider this moving aspect of his life to be under the eucalyptus shade of the environment *i.e.*, the unimagined force around him. This is the reason that never before the scientific and industrial revolution rather renaissance in Europe which took the world under its regime gradually, environment was not considered a problem by the mankind. In the course of industrial revolution the man became the master of so many scientific techniques and devices. Thus he was enabled to construct huge buildings, highways, bridges, rail-roads, mills, factories etc. to dig deep mines, to obtain coal, metals and minerals of so many kinds to establish large industrial sites. In the same order, he also turned to agricultural side of production by machines and manifestation of such mechanical and electrical appliances which caused him to get rid off hired immense human labour. Ultimately this procurement became the cause of environmental abuse, as the small towns swelled in the form of big cities and in the same time new cities were founded. Hence cultural disposition also took new turns. It was very prodigious involvement. Usually when problems emerge to be faced about, human mind begins to think to be rescued from them. But this problem was very immense. It had directly affected the cultural side of humanity. This caused the thinkers to add the cultural sphere also with the natural shpere of environment. Therefore, environment is now considered to be the embodiment of its branches *i.e.,* the natural environment and cultural environment. It has also been awarded a new name called Geographical environment.

In the wholesome conclusion, we can denote environment to be an embodiment of all the circumstances, effects and forces, which are vested in the social and cultural conditions, ready to affect the life of a person or a community. These circumstances can be brought forward by the nature and even by the mankind itself. Environment to its own factors on one side and is itself affected too on the other, as a whole.

When environment affects the cultural side of the social life, naturally it makes inroad into the manners, conduct, behaviour, thinking ambitions drawbacks, superstitions, traditions, customs and whatsoever is there not merely concerned to the activities of the society, but also concerned to its prevailing currents affected by its own other factors. This extends the boundaries of human knowledge which is offshooted in many channels and disciplines. The scientific thinking and attitude has overwhelmingly constituted the palace of knowledge very magnificent and majestic, which primarily it was considered to be related to the natural sciences only. But in the twentieth century, scientists marked the waves of industrial development swallowed dangerously causing environment to be of serious concern. This compelled to divert them towards the environmental research and they could realize the matter was of a serious nature. All the disciplines of knowledge were deeply related to environment. Biology, Geography, History, Economics, Sociology, Physics, Law, Political Science, Engineering, Botany etc. and none of the others too, was left by environment, untouched. Thus we find the environment now a matter of general consideration. Hence the vastness of its scope can be understood easily. Every government is bringing its best to make it a subject of household affairs and knowledge. That is why, it is an established fact that the scope of environment is not limited to studies only. We cannot exclude anything biological or non-biological character found on this earth, from the panorama of environment. Hence every discipline whether related to the biological studies *e.g.,* human beings, animals, plants and tiny creatures even, or to non-biological sphere of this creation have to associate with environment. Ultimately all the physical and social sciences have to analyze findings and theories in the light of the interacting responsive role of environment which prevail upon the all channels of social existence.

Prominently among all the organic sciences, biology, botany and bacteriology are apparently connected with environment. These are divided in many branches and sub-branches, which deal with so many specific classes and communities of the organic world. In physical sciences, physics, chemistry, mathematics, statistics, climatology, geography etc. have got deep affinity with environment. It is because the human life or the organic life as a whole even reacts in so many social, economical, cultural, historical, political and defensive manners. And hence all the social sciences and environment are deeply thick with each other. If the scholars study the life in view of the different aspects of physical and social sciences based on environmental geography, its scope can be distinguished to a considerable extent.

As it is established that environment is of multi-disciplinary nature, it has got very close relationship with the physical and other sciences. Herein after, the way in which environment is close to the other sciences is explained.

Biology and Environment. Biology is the composition of Zoology, Botany and Bacteriology. All these three sub-branches of biology have got the basic relations with environment. As it is known to us that it is only the human being who reacts expressively in the all possible ways. And tries to cope with it to any extent. Though such is the case with all the other organic world, but that is always defensive, thereby to save itself from the furious claws of environment while human being remain facing it and pollutes it too in usual course. But environment has its own system to regulate the world to which we can given an eye.

Zoology and Environment. Zoology studies the human being on one side and the creatures apart from the human world on the other. We study in Zoology the physical composition and activities of biomasse in relation to zoogeographic region. The creatures are divided in so many ways as for example, human, non-human, geographically surfacial and non surfacial in earth or into water. Moreover, the existence of different human species in different environmental situations—social and economical and physical formation, division of animal species according to their build habits etc. found in the different parts of the world all over and beneath geographical surface. Ecology is also studied in zoology, which is a part of environment.

Botany and Environment. Phytogeography is an important part of Botany. Studies in Phytogeography are related to the vegetable families which naturally exist and grow in the different geographical situations. From this point of view, the Earth is divided in four wide peninsulas of vegetation. These are—Arctic region, Northern temperate region, Tropical region, and Southern temperate region. Having in view, the kinds of vegetation found in these four zoogeographic regions, this has been categorized in twelve vast biomass which remain under direct influence of environment. This causes the environmental study in botany and geography both.

Bacteriology and Environment. There are certain principles according to which bacteria control system is evolved. We study these principles in bacteriology. It is the environment which causes the bacteria confined to certain areas. Therefore environment is deeply connected with geography and bacteriology.

Physics and Environment. Space (sky), the internal power of the earth solar energy, gravity etc. are the subjects of physics. The organism remains under influence of these powers but the medium of influence is environment. Analysis, calculation and study of all these topics have their concern to physics. Involvement of Geography is also there in the study of these topics. Hence physics, geography and environment are inter-related.

Chemistry and Environment. The physical atmosphere in which we live is constituted of so many gases when so, when the components of these gases are analyzed and the results derived as such along the methods adopted to take these in use, the subject deals with it all which is known by the name of chemistry. In the same way when so many things given by nature are subjected to such converted forms which are found prevalent in daily use, we must take it to be the work of chemistry. There is huge hoards of such things at present. Hence we can very easily understand that how far the industrialization would have been cause of polluting environment beacuse the production of these hoards cannot be had without releases of polluting nature. Beside this there are so many organic groups whose activities of living conditions are administered by chemical make up. Earth, water, air, heat, minerals are the subjects of chemistry. Therefore chemistry, environment and geography are inherently related.

Mathematics and Environment. Mathematics has become the driving force of research work of Environment. The developed devices are used to find out the exact figures and data in the course of research regarding environmental conditions of different nature as concerned to over and this populated areas, vegetation atmosphere, agricultural analysis of soil and so on. So many statistical methods and sampling techniques are used in this work. That is why mathematics is deeply to environmental statistics.

SOCIAL SCIENCES AND ENVIRONMENT

Due to so many reasons there are innumerable problems before this world—Political, social, economical etc. Social affairs conjoined to unfavourable environmental developments have seriously affected the mankind. Hence we have to study the matters concerned to the social sciences with a view of knowing, how these react under the influence of environment.

1. Sociology and Environment

Human being is known to be a social animal. Naturally it means that his development and progress has got the base in social environment. Social environment is one of the main features of environmental geography. Man is found in so many places *e.g.*, in villages, cities, towns and in open places too. Environment influences human living and forming social traditions and customs naturally differ from place to place. In other words, we can say that social behaviour of man is pertinent to his living place. Further the distinct social features are visible more prominently when we distinguish the human society as a whole in context to areas belonging to plains, mountains, river sides, sea shores, forests, valleys etc., or zoogeographic regions. Man born in different zoogeographic regions is quite distinguishable by his features even denoted to be of different species, The question speaks—why is it so ? The answer is there, that it all due to geographical environment spread over the different regions of this earth. This has given man, if we begin it with the age of his origin and end it with social structure, customs, behaviour and beliefs etc., a very remarkable distinctively. These are the basic functions, which divide man in different species. Hence we can very well conclude that socially man has got the original relation with the environment.

2. Economics and Environment

Economics is deeply and centrally related to production and distribution. Production is directly has its primary source in nature, wherever on this earth social pockets emerge, they struggle for survival. Naturally the produce of the land is found there, where these social pockets make their way to survive and their movements begin. These may be adopted by us in many ways to move according to their needs and procurement for a social make up and it is very possible that they might have been adopted in any of the ways we at present imagine, But whatsoever these might have been adopted they lead man to establish his base economically strong. Due to this reason the social pockets also move from one place to another, even to the extent of crossing zoogeographical regions. There might be so many other reasons of crossing of these regions but it is not impossible that the main factor might have remained economic. However, it is a well established fact that economic affairs are the most persuasive amongst all others. Historically we may locate it in the European continent. The Europeans crossed the oceans, which cause them to build empires even. The production is the next performance of the mankind. It is for man himself to organize his available means for economic progress. But these means are also the offshoots of environment. These differs in different geographical conditions. Usually geographical conditions are always there to make man dependant on them. Even the man's physical and psychological states are influenced by the environmental situations. From this point of view, we can compare the availability of resources in different zoo-geographical regions and the economic cultures developed there. Hence it is well proved that environment is directly connected to the progress of mankind economically. Ultimately economics has a deep concern to environment.

3. Political Science and Environment

Political science means the discipline which deals with the internal and external affairs of the kingdoms *i.e.,* the ruled territories by the legally formed government. It is quite apparent that so far as the internal affairs of the governments are concerned all of these have dignified their framework in the shapes of social organizations distinctive to one another. Hence each has developed its own living, thinking and working cultures. Naturally environment has remained conducive to such developments and the political backgrounds have come in formation. These are the environment circumstances which have remind dominating in respect of their struggle for their so many achievements on one side and losses too on the other. There are so many states, which contain social organizations of different cultures and beliefs even but their living cannot be opposite to the environment in which they live. However their traditions and beliefs become causes of internal struggle usually. These can be either of foreign nature or of the domestic one. Government have to deal with them accordingly. Naturally environmental conditions are there to be kept in consideration from the point of view of taking administrative measures to deal with that. Such is the case in dealing with international affairs. Generally environmental influences dominate the people adherent to foreign culture and beliefs, residing in the states of different cultures and beliefs. This situation create political problems. We can very well see this state of affairs in the political phases of India.

4. History and Environment

Miss Sepul has said *'History is geopraphy set into motion'*. And **Dr. K. M. Panniker** has written that *'Geography constitutes the permanent basis of every nation's history. It is not too much to say that it is one of the major factors that evolution of the people.'*

If we trace the history, not only of India but of all the countries of this world, we would apparently come to know that certainly the Governments play the key role in the formation of history, but the subsidiary factors, which are deeply related to the geographical conditions *e.g.*, physical environment constituted of surface of the earth, situations, span and climatic obstacles etc., are also there to play the lead role in deciding the future of the nations. It is the historical fact that the plains of Mesopotamia and Sindh played a great role in the histories of Middle Asia and India. Alexander could defeat much more strong army of king Puru due to rains.

In the same character environment influences social, political, historical and cultural sides of man. Hence disciplines concerned to the different aspects of the society are intimately related with the scope of environment. This facts is also helpful to understand the scope of environment.

ENVIRONMENTAL STUDIES

The modern age is the age of most developed civilization. In the known historical period of human remembarance, it seems that human being has reached to the climax of his achievement. Mental faculties of man have proved that he is certainly the master of all his survey. But in the wake of his astonishing discoveries he could not esteem the loss which he caused to this universe in the form of pollution of environment. However when the environmental balance began to disturb the human welfare the scientists could glimpse the injurious effects of strong industrialization. They became seriously attentive to divert themselves towards environmental relation of man with nature on one side and with the revolutionary progress of science on the other. But the scientists, could not come back now. They had carried the world to the atomic age. And now they are trying to lay foundations of cities and towns on moon, above in the sky and into the deep oceans. Naturally, in this age of developed science and technology everyone is supposed to know about the problem of environment; thereby to be aware of his duty as a citizen, not to do certain things, which are pollutative. Though every effort is being done to keep the environment free from defilement. For many of our social and political problems, it has become essential. That is why trained environmental biologists have been appointed to work in this sphere.

The fields of agriculture, forestry, fisheries, wild life management, space research, atomic research, public health, range management, animal husbandry, climatology, geology, geochemistry, and environmental physiology etc are the most important from the point of view of environmental study. Amongst the fields mentioned above public health is much more either neglected one or mismanaged one. The public at large does not feel abided to self-restraint from spoiling atmosphere. Hence the environmental problems are considerably serious not only in big cities and towns but even in small and big villages. It is possible that the inhabitants of these settlements face so many discrepancies too, which do not allow them to be too much careful about environment. Actually most of them understand environment beyond cleanliness. The same is the case with small-scale industries. Usually most of the small industrialists settle their industries in crowd pockets of the cities and the people living in the vicinity are unable to oppose such industrialists because they have certain type of relation with higher authorities and politicians.

Nowadays we find a crowded movement of petrol and diesel consuming vehicles in the cities. These are also one of the prominent sources of pollution.

Shortage of electricity is also a vice for the public health because the people have become more and more accustomed to its use but when the electric supply is detained, naturally the people begin to use generators which is a most terrible noise and smoke-producing source of pollution.

Though apparently it seems very difficult to get rid of pollution because it is directly concerned with the economic gains procuring devices and the social culture. If we take India for example, the state of affair is quite unsatisfactory. Merely appointments of environmental specialists cannot help much to fight with the environmental problems. Hence public awakening programmes must be instituted to make the people learn with basic principles of environment.

ATMOSPHERE

The cover of air that envelopes the Earth is known as the atmosphere. In other words, atmosphere is the protective thick gaseous mantle, surrounding the Earth, which sustains life of Earth and saves it from unfriendly environment of outer space. It extends to a height of about 1600 kilometers from the earth's surface.

Composition of Atmosphere

The atmosphere is composed of various gases and water vapour like—78.084% Nitrogen, 20.9476% Oxygen, 0.934% Argon, 0.0314% Carbon dioxide, 0.0018% Neon, 0.00052% Helium, 0.00015% Methane, 0.00011% Krypton, 5.0×10^{-5}% Hydrogen, 2.0×10^{-5}% Nitrous oxide, 1.2% Carbon monoxide, 8.0×10^{-6}% Xenon, 2.0×10^{-6}% Ozone, 6.0×10^{-7}% Ammonia, 1.0×10^{-7}% Nitrogen dioxide, 6.0% Nitric oxide, 2.0×10^{-8}% Sulphur dioxide, 2.0×10^{-8}% Hydrogen sulphide.

Structure of Atmosphere

Atmospheric phenomena associated with the earth's magnetic and gravitational field extend outward for several thousand kilometres to a vague zone of nebulous gases and radiation particles that become rarer and rarer until at least characteristics of the atmospheric cases.

Structure of Atmosphere along with Temperature

On the basis of temperature profile and other related phenomena, atmosphere is divided into four major layers:

1. **Troposphere**
2. **Stratosphere**

3. Mesosphere

4. Thermosphere

(*i*) *Troposphere.* It is the lower portion of the atmosphere, extending upto about 8 km at the poles and 16 km at the equator.

(*ii*) *Stratosphere.* Above the troposphere lies the stratosphere, where temperature is nearly constant upward to about 20 km and then increases upto a maximum of 0° C near its outer limit.

(*iii*) *Mesosphere.* The temperature decreases slowly with the attitude but then sharply to a minimum of about near the mesopause, at 80 km. Most burn and disintegrate as they experience increasing friction, in this layer.

(*iv*) *Thermosphere.* The temperature again rises to very high values and at times approaches 2,000° C, and even more at about 500 km, depending upon solar activity.

Table 1.1. Principal Gases of Dry-Air in the Lower Atmosphere

S.No.	*Gas*	*Concentration % by Volume*	*Concentation PPm by Volume*
1.	Nitrogen (N_2)	78.084	7,80,000
2.	Oxygen (O_2)	20.9476	2,09,500
3.	Argon (Ar)	0.934	9,340
4.	Carbon dioxide (CO_2)	0.0314	314
5.	Neon (Ne)	0.0018	18
6.	Helium (He)	0.00052	5.2
7.	Methane (CH_4)	0.00015	1.5
8.	Krypton (Kr)	0.00011	1.1
9.	Hydrogen (H_2)	5.0×10^{-5}	0.5
10.	Nitrous oxide (N_2O)	2.0×10^{-5}	0.2
11.	Carbon monoxide (CO)	1.2×10^{-5}	0.12
12.	Xenon (Xe)	8.0×10^{-6}	0.08
13.	Ozone (O_3)	2.0×10^{-6}	0.02
14.	Ammonia (NH_3)	6.0×10^{-7}	0.006
15.	Nitrogen dioxide (NO_2)	1.0×10^{-7}	0.001
16.	Nitric oxide (NO)	6.0×10^{-8}	0.0006
17.	Sulphur dioxide (SO_2)	2.0×10^{-8}	0.0002
18.	Hydrogen sulphide (H_2S)	2.0×10^{-8}	0.002

HYDROSPHERE

It covers more than 75% of the earth's surface either as oceans or as fresh water. Hydrosphere includes sea, rivers, oceans, lakes, ponds, streams etc. Most of the earth's surface water is in the oceans which about 35 parts per thousand of dissolved salt. Of the remainder, most of fresh water with a salt content of 0.2% is found either in lakes and ponds or in river and stream, fresh water is also available in from of rains, snow, dew etc. Evaporation of water from oceans, cloud formation and precipitation are responsible for world wide supply through hydrologic cycle.

Composition of Natural Water

The pure water is the one which contains two parts of hydrogen and one part of oxygen by volume, and by weight, it consists of one part of hydrogen and eight part of oxygen. It possess abnormal physical properties compare to hydrogen sulphide (H_2S), hydrogen selenide (H_2Te). It is a liquid at room temperature with melting point 0°C, boiling point 100°C, density 1.0 high dipole moment, 1.84×10^{-3} high dielectric constant, 80 degree specific heat, 1.0 cal per g per degree C, high heat of evaporation 540 cal/g, surface tension 73 dynes/cm at 20°, and viscosity 0.01 poise at 20°C. These abnormal properties of water are due to hydrogen bounding. The structure of water as shown by x-ray diffraction studies on ice crystal shows that in water there exists an arrangement in which each water molecules is surrounded by four neighbouring water molecules.

Oceans cover about 70.8% of the earth's total surface area and hold more than 97% of the earth's water supply. This water despite its abundance is not directly useful to man, it contains about 2,000 times more dissolved salts than fresh water and has pH of about 8.1 (± 0.2) all over the globe. The most common dissolved salts are chloride and sulphate of sodium, magnesium, calcium, potassium and strontium. The ocean water contain about 3.5% of dissolved salts. Chemically speaking, ocean water is a solution of 0.5 M NaCl and 0.05 M $MgSO_4$, containing traces of all conceivable matter in the universe. The most abundant element in sea water are chlorine = 19×10^{-3}, sodium = 10.5×10^{-3}, magnesium = 1.35×10^{-3}, sulphur = 885, calcium = 400, potassium = 380, bromine = 65, carbon = 28, strontium = 8, boron = 4.6 and in river and lake water are—carbonate = 35.2, sulphate = 12.1, chloride = 5.7, nitrate = 0.9, silica = 11.7, calcium = 20.4, sodium = 5.8, magnesium = 3.4, potassium = 2.1.

Table 1.2. Typical Quality of Natural Water

S.No.	*Quality Parameter*	*Average Normal Value and Quantity*
1.	pH value	6.8–7.0
2.	Hardness ($CaCO_3$)	19 mg/L
3.	Presence of Calcium ($CaCO_3$)	16 mg/L
4.	Presence of bicarbonate ($CaCO_3$)	12 mg/L
5.	Presence of sulphate (SO_4)	10 mg/L
6.	Presence of chloride (Cl)	9 mg/L
7.	Presence of sodium (Na)	6 mg/L
8.	Acidity	4 mg/L
9.	Presence of magnesium (Mg)	3 mg/L
10.	Presence of ammonia (NH_3)	0.8 mg/L
11.	Presence of nitrate (N)	0.1 mg/L

Groundwater generally, is clear but harder than the surface water of the region in which it occurs, and its quality tends to be uniformed throughout an aquifers a part of the surface water infiltrates into the ground, it becomes comparatively free from most of the impurities present in surface water, but it may contain dissolved impurities like iron, magnesium, calcium, chloride, fluoride, dissolved gases such as oxygen, hydrogen, carbon dioxide, methane etc., depending on chemical characteristics of the water and soil of the aquifer recharge area.

LITHOSPHERE

Lithosphere means the mantle of rocks constituting the earth's crust. The earth is a cold spherical solid planet of solar system, which spins on axis and revolves around the Sun at a certain constant distance. The solid component of the earth is called lithosphere, which includes mainly soil, earth, rocks, and mountains etc. The lithosphere mainly contains three layers—Crust, Mantle and outer and inner Core. The core is the central fluid or vapourized sphere having diameter of about 2,500 km from the center. Core is probably composed of nickel and iron. The mantle extends about 2,900–3,000 km above the core and is in the solid state. The outermost solid zone of the earth is known as crust. It is about 8.40 km above the mantle. The surface of crust is covered with the soil.

Table 1.3. Physical Characteristics of Earth

Zone	*Volume* (× 10^{-27} cm^3)	*Thickness* (km)	*Mass* (× 10^{27})	*Mean Density* (gm^{-3})	*Mass* (%)
Crust	0.008	17	0.024	2.8	0.4
Mantle	0.899	2,883	4.075	4.5	68.1
Core	0.175	3,471	1.876	10.7	31.5
Whole Earth	1.082	6,371	5.975	5.5	100.0

BIOSPHERE

Biosphere indicates domain of living organisms and their interactions with environment, *i.e.*, the atmosphere, hydrosphere and lithosphere. The planet Earth along with its living organisms and environment which sustains life is known as biosphere. The biosphere extends vertically into the atmosphere to about 10 km, downward into the ocean to a depth of about 35,000 Ft and about 23,000 Ft into the earth's surface itself where life has existed. Green plants through photosynthesis accumulate life supportings in the atmosphere, animal inhale oxygen during respiration and given out carbon dioxide which is utilized by plants during photosynthesis.

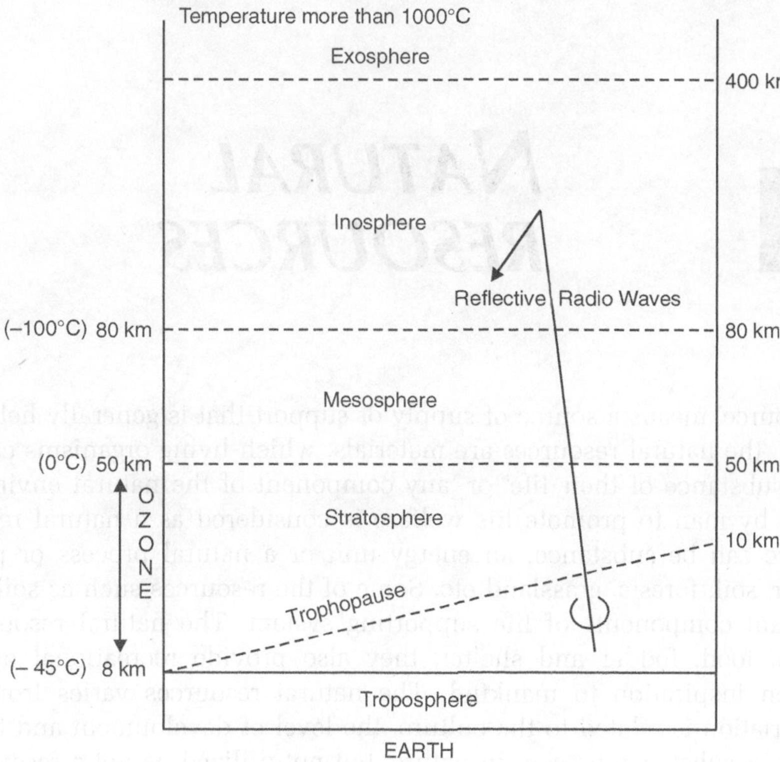

Fig. 1.1.

EXERCISES

1. What is the meaning of Environment ? Discuss its scope and importance in short ?
2. How environmental studies is beneficial to us ?
3. Write short notes on:
   (*a*) Hydrosphere
   (*b*) Atmosphere
   (*c*) Lithosphere
   (*d*) Biosphere

□□□

Unit II NATURAL RESOURCES

The world 'resource' means a source of supply or support that is generally held in reserve. In other words, 'the natural resources are materials, which living organisms can take from nature for the substance of their life' or 'any component of the natural environment that can be utilized by man to promote his welfare is considered as a natural resource'. The natural resource can be substance, an energy unit or a natural process or phenomenon *e.g.*, land, water, soil, forests, grassland etc. Some of the resources such as soil, water, etc., are the important components of life supporting system. The natural resources are not only sources of food, fodder and shelter, they also provide recreational opportunities, solace and even inspiration to mankind. The natural resources varies from society to society. The variation is related to the culture, the level of development and the nature of work of society. A substance present in nature, but not utilized, is not a resource *e.g.*, one tribal society of the Andman group of islands do not use uranium, gold and silver, which are therefore, not resources for them. Similarly, uranium was not of much importance earlier but has now become valuable resource because of its use in nuclear energy.

CLASSIFICATION OF NATURAL RESOURCES

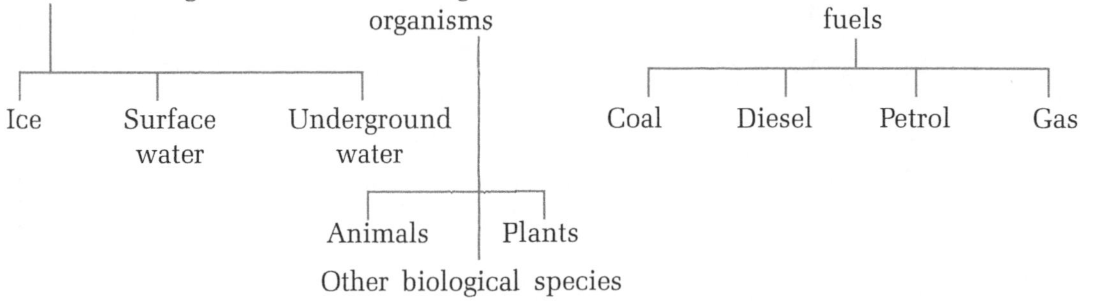

1. Inexhaustible Resources

These resources are viable in abundance in the nature and they are not likely to be exhausted by human activities. Some inexhaustible resources remain unaffected by human activities, while many other may show some changes in their quality. Solar energy, wind power, rainfall and even atomic energy are classified as inexhaustible resources. These resources cannot be exhausted significantly at global level due to human activities. Some resources may sometimes be affected locally by human activities *e.g.*, the quality of air is changed due to pollution.

2. Exhaustible Resources

The resources have limited supply on the earth, and are, therefore, liable to be exhausted if used indiscriminately. Exhaustible resources are of two types—*renewable* and *non-renewable*.

(*i*) *Renewable Resources*. These resources have the capacity to reappear or replenish themselves by quick recycling, reproduction and replacement within a reasonable time. Soil, water and living things *i.e.*, plants, animals and micro-organisms are renewable resources. The growth and reproduction of living beings can be successful managed so that these resources are continuously regenerated. However, if the consumption of these resources contains to exceed their rate of renewal not only their quality becomes affected, they may even totally exhausted.

(*ii*) *Non-renewable Resources*. These resources lack the ability of recycling and replacement. The substances with a very long recycling time are also regarded as non-renewable resources.

Many abiotic resources are non-renewable. The fossil fuels (coal, petroleum and gas) and metals once extracted cannot be regenerated at the place of extraction. After unlimited extraction and use, the fossil fuels will definitely get exhausted. Some biotic resources are also non-renewable. The biological species, which have evolved in nature during the coarse millions years are considered non-renewable. A biological species which becomes extinct from the earth cannot be created again.

Table 2.1. Important Ecosystem and their Renewable Products

Forest	Yield Timber and other plant products
Range Lands (Grasslands)	Sustain grazing animals for milk, wool and meat products.
Wildlife	Maintain food chain
Agricultural Systems	Yield food and fibre
Marine and Fresh Water System	Yield various foods from plants and animals

PRINCIPAL NATURAL RESOURCES AND THEIR PROBLEMS

Forest Resources

A forest is a biotic community which is predominantly composed of trees, shrubs or any woody vegetation and often with a closed canopy (overhanging covering). Approximately one-third of the earth's total land area is covered by forests. Forests are the valuable wealth of a country. They are storehouse of biodiversity and provide important environmental services to mankind. To offer these services, forests have following three type of functions:

(i) *Productive Functions.* The productive functions of forests include the production of timer, bamboo's, food and a wide variety of compounds, such as resins, alkaloids, essential oils, latex, pharmaceuticals, etc.

(ii) *Protective Functions.* These functions include conservation of soil and water, prevention of drought and protection against wind, cold, radiation, noise, sights and smells, etc.

(iii) *Regulative Functions.* The regulative functions involve absorption, storage and release of gases (like CO_2 and O_2), water, mineral elements and radiant energy. Floods, draughts and global biogeochemicals cycles, particularly of carbon is also regulated by forests. The regulative functions of forests improve atmosphere and temperature conditions, and enhance the economic and environmental value of the landscape.

Forest Area in India

India had abundant forests in ancient times. Chandragupta Maurya and Kautilya took keen interest in forest management. Ashoka's stone edicts prescribed tree planting. But increase in population, lack of foresight and unplanned felling of trees have drastically reduced the area and quality of forests in India today. At the beginning of the 20th century about 30% of land in India was covered with forests. But by the end of 20th century, the forest cover was reduced to 19.4%. This is far less than the optimum area of forest recommended by the **National Forest Policy** (1988). The National Forest Policy has recommended 33% forest area for the plains and 67% for the hills. Of the existing forests, less than two third are dense forests, and the rest are open degraded forests. In India; I, the per capita forest area available for below the average for the world. Today per capita forest area available in India as 0.06 ha (hactre) as against 0.64 ha of the world's per capita forest area.

Forest cover in India according to 1999 estimate is given in the following table:

Table 2.2.

Description	*Area (sq.km)*	*% age Geographical Area*
Dense forest (Canopy cover > 40% of land)	3,77,358	11.5
Open forest (Canopy cover 10.40% of land)	2,55,064	7.8
Mangrove forest (Canopy cover < 20% of land)	4,871	0.1
Scrub (Canopy cover < 20% of land)	5,896	1.6
Non forest (Other land use)	25,98,074	79.0
Total	32,87,263	100.0

Uses and Over-Exploitation

Following are the uses of forests:

1. **Check air and noise pollution.** Forests can absorb many toxic gases and these keeps the air pure. Trees can also absorb noise and lower down the noise pollution level.
2. **Reduce wind erosion.** In deserts, forest reduce wind erosion by checking wind velocity.
3. **Reduce soil erosion.** Forests bind the soil particles with root and prevent soil erosion, particularly in hilly areas.
4. **Prevent landslides.** Forests also reduce landslides by binding the soil particles with their roots.
5. **Reduce global warming.** Trees absorb carbon dioxide and release oxygen during photosynthesis thereby reducing carbon dioxide concentration in the atmosphere.
6. **Regulation of hydrological cycle.** Forested water shed act like giansponges, absorbing the rainfall, slowing down the runoff and slowly releasing the water for recharged of springs. About 50–80% of the moisture in the air above tropical forests comes from their transpiration and helps in bringing rains.
7. **Commercial uses.** Trees mainly provide wood which is used as fuel wood and also used as a raw material for various industries as pulp, paper newsprint, board, timer for furniture items. It is used for packing material, matches, sport goods, etc.
8. **Protection to wildlife.** Forests provide homes to millions of wild animals and plants.
9. **Aesthetic-uses.** Trees also play a dominant role in the beautification of the environment.
10. **Production of oxygen.** The trees produce oxygen by photosynthesis and are also known as earth's lungs.
11. **Reduce flood havoc.** Trees reduce the flood havoc by absorbing the excess water.
12. **Balancing the gaseous cycles.** Forests balance the carbon dioxide and oxygen ratio in the atmosphere.
13. **Recreational uses.** With the development of national parks and sanctuaries, trees have a role in recreation too.

Exploitation. Forests have served mankind the very beginning of its history. Now, with growing population, the demand for raw material like timber, fuel, wood has increased and resulted in large scale wood felling, destruction, road, building and clearing of forest also contribute to the national economy. The international timber trade above is about US $ 40 billion per years. But the excessive use of fuel wood and charcoal, expansion of urban, agricultural and industrial areas, overgrazing, deforestation, desertification, have together led to exploitation of forest.

DEFORESTATION

Destruction of forest is a formidable threat to the quality of life, country's economy and future development. World's forest cover has been shrinking rapidly, especially in the developing countries located in tropics. While the temperate forests have lost only 1% or less of its area, the tropics have lost more than 40% of the forest cover due to deforestation. The current rate of deforestation is estimated to be more than 10 million ha per year. If this rate of deforestation continues, it is feared that the remaining tropical forests may disappear within a century.

Causes of Deforestation

The main causes of deforestation are as follows:

(*i*) *Explosion of Human Population.* Due to over-population of human beings, requirement of timber, fuel, wood, paper etc., has been increased. Man has cleared large areas of forest for agriculture, housing, factories, roads and railway tracks.

(*ii*) *Shifting Cultivation (Jhum).* Shifting cultivation or Jhum is traditional agro-forestry system widely practiced in the north eastern region of our country. It involves felling and burning of forests followed by cultivation of crops for few years, and abandoning cultivation to allow forest's regrowth. This type of cultivation is always meant to fulfil local needs but this method causes extreme damage to forests.

(*iii*) *Construction of Roads.* Construction of roads along the mountains which cover nearly 30,000 km in ecologically fragile area is another cause of forest degradation.

(*iv*) *Demand of Wood for Industries.* Wood is used for several industrial processes, such as making boxes, crates, packing cases, furniture, match boxes, paper, plywood etc. The requirement of wood for industries has increased tremendously in the country in the recent years. Today, the country's annual requirement for industrial wood is about 40 M3m (Mega cubic metre), whereas only 13M3m is available.

(*v*) *Mining Operations.* Mining operations have a serious impact on forest areas. Large areas of forest are cleared and laid barren as result of open cast mining of mica, coal, manganese, limestone, etc. The result is turning of forest into pastures.

(*vi*) *Overgrazing.* Overgrazing of forests by livestock has resulted in large scale degradation of forests.

(*vii*) *Pests.* Many kinds of insects and pests destroy trees by eating up leaves, boring into shoots and spreading diseases.

(*viii*) *Fires.* Fire is the worst enemy of forest. It destroys the full grown trees, seedlings, seeds and even humus. It also causes large scale damage to animal life.

(*ix*) *Weather.* Extreme weather conditions such as frost, storms and heat also destroy forests.

(x) *Dams and Hydroelectric Projects.* Dams reservoirs and hydroelectric projects submerge forest, displace local people cause water logging and siltation and may result in earthquakes. Such projects displace large number of tribal people and cause great damage to the forests, which are the main source of their livelihood. The relief and rehabilitation programmes offered by the government always seems inadequate.

Evil Effects of Deforestation

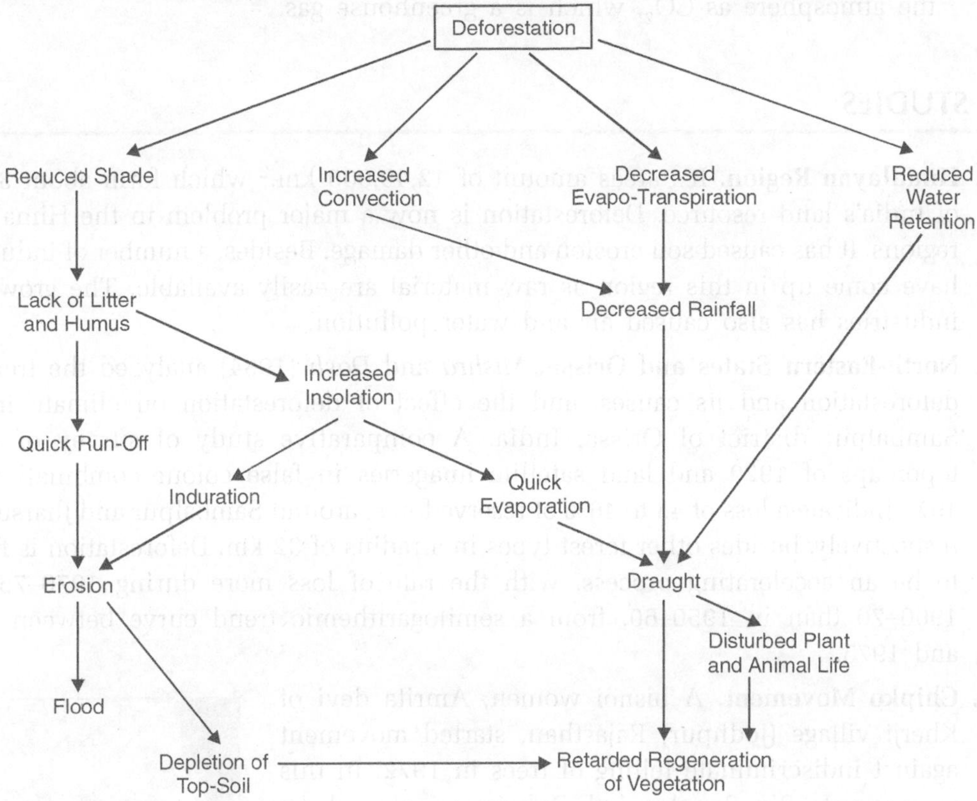

Fig. 2.1. *Effects of Deforestation*

Deforestation affects human life in several ways. The main effects of deforestation are as follows:

(*i*) Deforestation results in increased soil erosion and decreased soil fertility. In drier area, deforestation can lead to the formation of deserts.

(*ii*) Forests, particularly on mountains, provide considerable protection from floods by trapping and absorbing precipitation *i.e.*, rains, snow etc., and slowly releasing it later. When the forest is removed, the amount of runoff water flowing into river and streams increases several folds causing frequent floods.

(*iii*) Human beings are deprived of benefits of forest trees and wild animals.

(*iv*) Deforestation causes the extinction of forest dwelling plant, animal and microbial species resulting into a loss of irreplaceable genetic resources.

(*v*) Deforestation also threatens indigenous (tribal) people whose culture and physical survival depend upon the forests.

(*vi*) The pattern of rainfall has changed in deforestation areas. The rainfall has declined and droughts have become common.

(*vii*) Deforestation induces regional and global climate change. The climate has become warmer due to lack of humidity in the deforested regions.

(*viii*) Deforestation also contributes to global warming by releasing stored carbon into the atmosphere as CO_2, which is a greenhouse gas.

CASE STUDIES

1. **Himalayan Region.** It's areas amount of 12,49,000 km.^{2} which form about 38.4% of India's land resource. Deforestation is now a major problem in the Himalayan regions. It has caused soil erosion and other damage. Besides, a number of industries have come up in this region as raw material are easily available. The growth of industries has also caused air and water pollution.

2. **North-Eastern States and Orissa.** *Mishra* and *Dash* (1984) analyzed the trend of deforestation and its causes, and the effect of deforestation on climate in the Sambalpur district of Orissa, India. A comparative study of survey of India topomaps of 1929 and land satellite imageries in false colour combinations of 1975 indicate a loss of 43 to 46% of reserve forest around Sambalpur and Jharsuguda respectively, besides other forest types in a radius of 32 km. Deforestation is found to be an accelerating process, with the rate of loss more during 1970–75 and 1960–70 than in 1950–60, from a semilogarithemic trend curve between 1929 and 1975.

3. **Chipko Movement.** A bisnoi women, Amrita devi of Kherji village (Jodhpur) Rajasthan, started movement against indiscriminate felling of trees in 1972. In this movement, Sir Sundar Lal Bahuguna played an important role. He started peace march of 3000 km against the felling of trees. Chipko movement was started in 1972 in Tihri-garhwal, U.P. (at present Uttranchal). In the beginning menfolk never got success. Then, womenfolk started movement with menfolk and hugged the trees and saved from cutting by the axed man of the contractor. In Karnataka, Chipko movement is known as Appiko movement. It has three objectives:

Sir Sunderlal Bahuguna

   (*i*) Ulisu ⇒ To conserve

   (*ii*) Belesu ⇒ To grow

   (*iii*) Balaru ⇒ Rational use

Following were the outcomes of the movement:

1. The U.P. Government abolished the contractor system of tree felling.
2. It stopped the process of deforestation.
3. Forest Development Corporation was established for welfare of hilly human being.
4. It inlightned the society about the importance of ecological balance in nature.

TIMBER EXTRACTION

Economic Basis of Timber Extraction. Forests gives us important products to economies. Wood is primary forest product. Wood is light weighted, durable and waterproof material. The extraction of wood requires less investment and the wood is usually of high value. The old growth timber cut for engineering purposes like building house, making, furniture etc. Extraction of timer has a well known effects. Removal of forests produces a loss of animal habitat and loss of biodiversity. These effects are directly related to well-being of human population in deforested areas. Their primary lifestyle are force based. Expanded civilization has a negative impact on natural forests.

MINING

Mining Effects on Environment and Human Health

Mining process has adverse effects on environment, which are given below:

1. Land damage results from the underground mining.
2. Removal of vegetation with underlying soil mantle and overlying rock masses due to mining process leads to defacing the topography and destruction of the landscape in the area.
3. Wastes from uranium mines leads to air pollution in radioactive radiation's.
4. Mining for minerals also disturb the soil chemistry through turning soil profiles upside down.
5. Discharge of salts and acids from coal mines produces water pollution.
6. When forest cover declined and non-forest area increases. Due to this unstable zones are developed which leads to landslides.
7. Ecological problems evolved when native metals extracted from streambeds.
8. Mining process produces health hazards and accidents, resulting in several deaths and causalities.
9. From the underground mining, Pheumoconiosis disease is commonly developed. When coal dust inhaled by the miners impairs respiration by damaging the lung tissues, leads to disability and death results when no cure.
10. Chasnala mines tragedy in Bihar is famous and results from excessive mining of coal.

DAMS AND THEIR EFFECTS ON FOREST AND TRIBAL PEOPLE

Impact of Dams on Forests and Tribal People

Forests are directly and indirectly affected by dams. Hydro-electric dams are main cause for deforestation. About 40,000 large dams are currently obstructing the workloads rivers. Destruction of forests occur for constructing big dams, which alters the natural ecological balance. In this way landslides, droughts, and floods conditions may be arised in such area. Socio-economic problems related to tribal or native people results from big dam construction.

Dam construction produces a number of health hazards. Thousands of workers who build the dam caught by the diseases like AIDS, measles, tuberculosis, syphilis etc. Dam-building has resulted in wide range human rights violations. Rehabilitation policy of the government is important and typical when most of the displaced persons are tribal people. Tribal life and culture are widely associated with forests.

WATER RESOURCES

Water covered 71% of earth's surface. Total volume of water in hydrosphere 97% is ocean water and rest 3% is available as fresh water. The total volume of groundwater found in underground reservoirs, called as aquifers. Agriculture demands the maximum amount of water in the world. But excessive use of groundwater depletes aquifers, leads to lowering of water-table and produce salinisation, water logging and alkalinisation of soils. Todays, population of the world required 6 billion m^3 or 6 km^3 water per year.

The distribution of fresh water is not uniform and the mode of use also varies. In this way, there is a scarcity of fresh water in many parts of the world.

Type of Water Resources

There are different types of water resources found in nature as water occurs on the land, below its surface, in atmosphere and in the biomass. About 97% water present in the vast oceans, 2% packed in the form of ice-sheets and less than 1% is available as fresh water. This fresh water comes from inland surface water sources such as rivers, streams, lakes and reservoirs and groundwater source.

Various Form of Water Resources

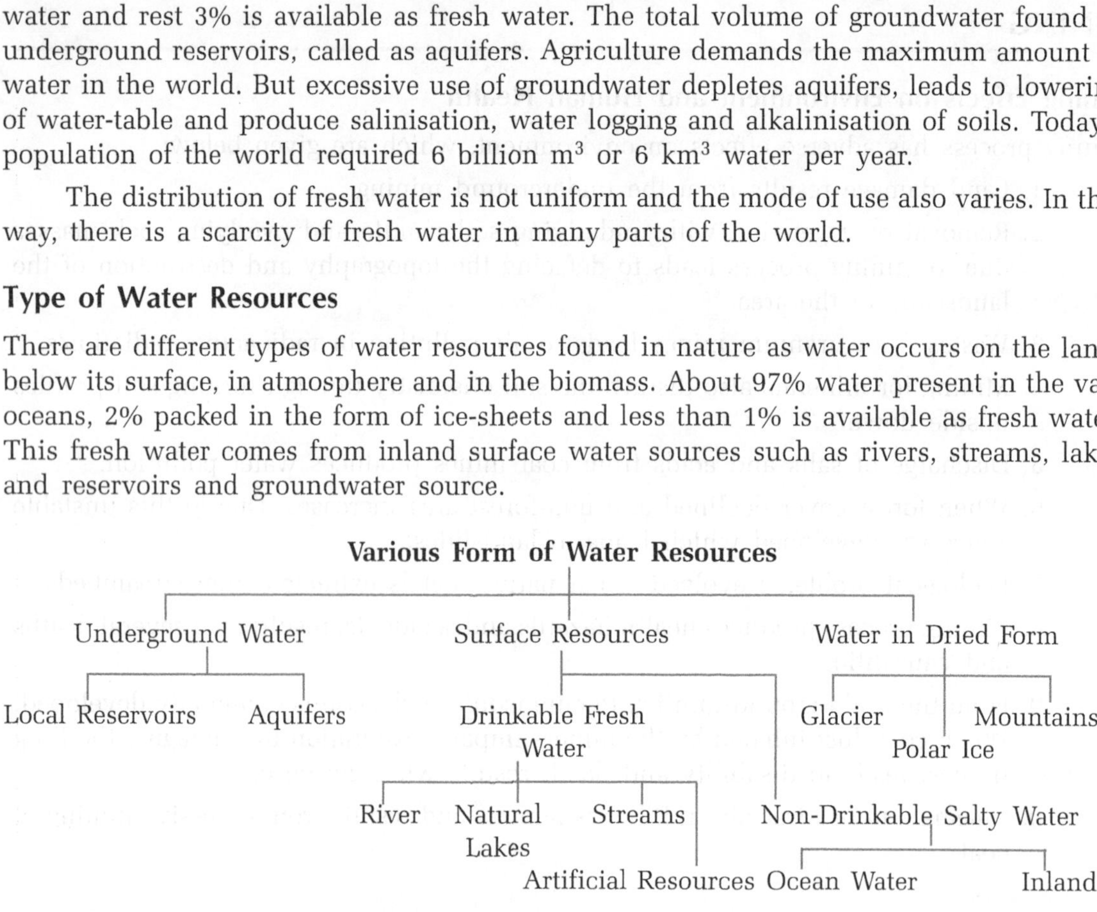

1. Surface Water Sources

Surface water generated directly from precipitation in the form of rainfall or snow on land surface as well as in oceans.

Surface water on the land is the main source of water for human needs like agriculture and industries. The rate of surface water mainly depends on total precipitation and its seasonal distribution. There are different types of surface water sources given below:

(*i*) *Rivers and Streams.* Rivers and streams are the main source of water supply. Water from such sources has variable quality and less satisfactory than the water from lakes and impounded reservoirs.

(*ii*) *Natural Lakes and Ponds.* Lakes are important source of fresh water, but ponds are temperate or permanent shallow water bodies. In this way they are known as an important source of fresh water supply.

(*iii*) *Sea Water or Oceans* consists of 97% of the total volume of water but ocean water characterized by salinity and temperature. Salinity is due to presence of number of dissolved salts. Average salinity of sea water is about 35 per thousand or 35%, so it becomes uneconomical.

2. Groundwater Sources

Groundwater is the other source of fresh water reservoir. Rain water or snow-melt that neither runs off the surface nor evaporates but sinks into the ground is called as underground water'. Finally groundwater reaches the sea through underground channels. So, it makes a part of the hydrological cycle. When the groundwater withdrawn rate is higher than the rate of natural recharge it may leads to several kinds of problems such as

1. Water table lower, when the rate of pumping is higher, so it allow wells to go dry or can deplete a full aquifer.
2. Highest pumping rate causes pores formations to collapse, resulting in subsidence or settling of the above surface.

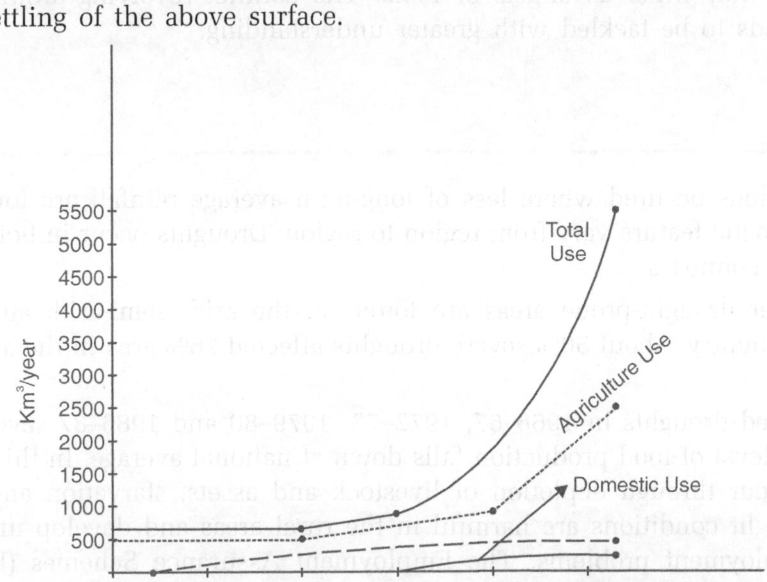

Fig. 2.2. *Expected Increase in Total Water use during the 20th Century.*

Uses of Water Sources

The rate of total water use three times faster than the population growth rate. In near future, doubling of the world's population mean a six fold increase in the total global water requirements of the total water used in India, 92% is for irrigation and 8% for industrial and domestic use. Scientists at the Indian Institute of Technology in New Delhi measure and ultimate usable potential of 86.5 mhm. The utilizable potential of groundwater is estimated to be around 35.58 mhm out of about 50 mhm.

Over utilization of surface and groundwater. The total consumption of water in 1974 was 38 mhm, but in 2025 it reaches the maximum utilizable potential. In this way we find that by 2025 if we consume more surface water than its utilizable potential, the consumption of surface water is 70.88 mhm by 2025 while maximum utilizable potential is 68.41 mh. However, groundwater consumption would be about 70% of utilizable potential by 2025.

Agriculture is a major field of consumption of surface water. In 1985, we used 17.40 million hectare metre (mhm) groundwater and most of it used for irrigation purposes only.

Conflicts over water. Millions of years ago life evloved in water and water became a basic unit of every living being. However, in most parts of our country, water scarce. When water is scarce, people should understand its value and should not waste it. Rapid population growth, industrial expansion and urbans migration are threatening India's water resources.

The two South Indian states, Tamil Nadu and Karnataka, have been bitterly squabbling over sharing of the Cauvery waters. Water tankers supplying drinking water in many parts of the country are a common sight. Conflicts become acute in the southern parts of India. The general idea is to transfer water from the Brahmaputra and the Ganga systems westwards to southern Uttar Pradesh, Haryana, Punjab and Rajasthan and southwards to peninsular component. Conflicts may also produce when water systems are used as instruments of war, either as targets or tools. The conflict revolving around sharing of river water needs to be tackled with greater understanding.

DROUGHTS

Drought conditions occured where less of long-term average rainfall are found. Drought is a natural climatic feature vary from region to region. Droughts occur in both developing and developed countries.

Some other drought-prone areas are found in the arid, semi-arid, and sub-humid regions of the country. About 50% severe droughts affected 76% area in the arid and semi-arid zones.

Indian faced droughts in 1966–67, 1972–73, 1979–80 and 1986–87 severely. In each conditions the level of food production falls down at national average. In this way, a large scare losses occur through depletion of livestock and assets, starvation and high range mortality. Drought conditions are harmful in the rural areas and develop unemployment and under-employment problems. The Employment Assurance Schemes (EAS) provide employment opportunities in drought-prone areas. There are three types of droughts, which are inter-related in several ways.

Types of Drought

1. **Hydrological Drought.** Hydrological drought is related with the effects of periods of precipitation short falls on surface or sub-surface water supply.

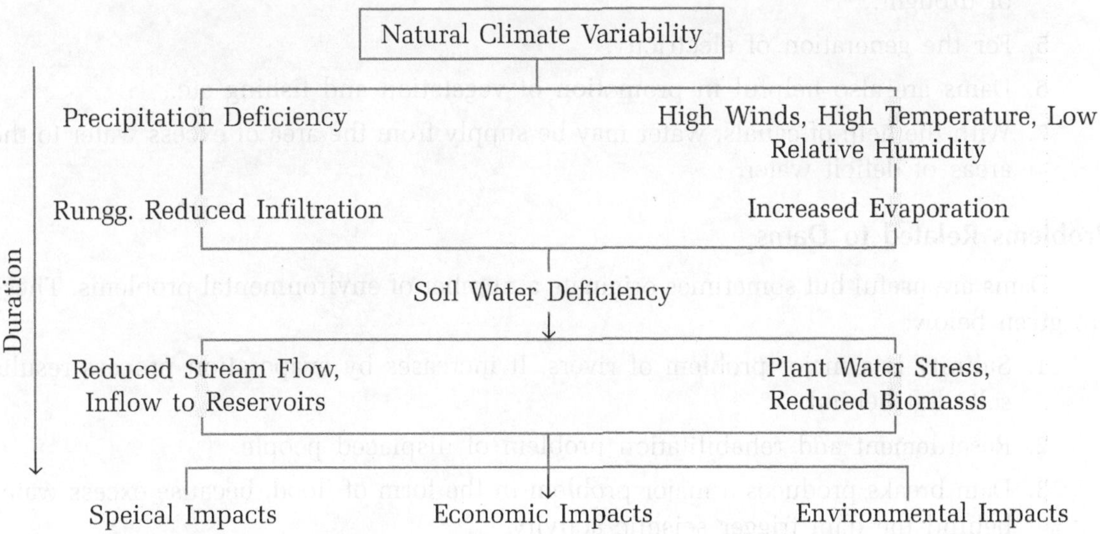

2. **Meteorological Drought.** This type is usually depends on the degree of dryness and the duration of the dry period.
3. **Agricultural Drought.** Agricultural drought is mainly related to meteorological drought to agricultural impacts, occurring or reduced groundwater levels, rainfall shortages, difference between actual and potential evapo-transpiration and so on.

FLOODS

Floods occur whenever the magnitude of low water exceeds the carrying capacity of water channels. This has a ravaging effect on the surroundings, which produces enormous loss of agricultural crops, livestock and human life. Floods are not always disastrous. For instances, annual floods provide fertile food plains, suitable for crop production.

India is the most flood affected nation after Bangladesh. Every year, with unfailing regularity, the monsoons occur, which bring destruction, despair and one fifth of global deaths on account of floods.

DAMS—BENEFITS AND PROBLEMS

Benefits of Dam

The first dam was built in 1890. Dam are trap run off and storage reservoirs so as to impound hundred amount of rain water. This collected water used for irrigation, water transport etc. Benefits of dam are given as follow:

1. Dams are used for supplying drinking water.

2. For saving the land area and protection of soil.
3. Flood conditions controlled by dams.
4. Dams supply water for irrigation purpose during the dry conditions in the form of drought.
5. For the generation of electricity.
6. Dams are also helpful in promotion of vegetation and fishing etc.
7. With the help of canals, water may be supply from the area of excess water to the areas of deficit water.

Problems Related to Dams

Dams are useful but sometimes originate a number of environmental problems. These are given below:

1. Salinity is a major problem of rivers. It increases by evaporation process results salts left behind.
2. Resettlement and rehabilitation problem of displaced people.
3. Dam breaks produces a major problem in the form of flood, because excess water behind the dam trigger seismic activity.
4. A large area covered by dams include fertile fields and human settlements.
5. Population living near the dam spill, faced a number of water wasted disease like malaria.
6. Accumulation of sediments makes dams useless and loss of suitable nutrients to the agricultural lands.

MINERAL RESOURCES

Mineral are finite and non-renewable resource. "All organic and inorganic substances comes under mineral resources, these are derived from the earth's dust". Some minerals consists of a single element such as gold, silver, diamond and sulphur. Minerals usually occur as ores, combined with many other substances. Our industries requires a number of minerals like copper, coal, petrol etc. Ores are first processed and then minerals are extracted. Some industries depends on use of petroleum, natural gas and coal. Minerals demand increases day by day. Industries self depends on about 80 of the known minerals.

Types of Mineral Resources

Minerals are mainly divided into three categories as fuel, metallic and non-metallic. Fuel minerals are coal, petroleum, natural gas. Metallic minerals are iron ore, chromium ore, bauxite etc. Non-metallic minerals are clay, dolomite, lime stone etc. In whole production of minerals about 88% covers by fuel minerals and rest covers by the metallic and non-metallic minerals.

(A) Fuel Minerals

Fossil fuel consist of coal, oil and natural gas. Our country rich in coal reserves but poor in oil and natural reserves.

(*i*) *Coal.* Coal is the best source to synthesize number of chemistry compounds. Major states known for the coal reserves are Jharkhand, some parts of Bihar, Orissa, West Bengal, Maharashtra. Coal typically contains 3.4% of sulphur which comes SO_2 during combustion, which sometimes combines was moisture forming H_2SO_4 acid and settles as acid rain.

(*ii*) *Crude Oil (Petroleum).* India is an oil poor country. Intensive exploration in potential area indicated sources with an annual production rate of 227 million barrels. There are about 40,000 fields but only 10% of them have a significant impact on the world oil production. In India, oil was first found at Makum but drilling oil started at Digboi in Lakhimpur district.

(*iii*) *Natural Gas.* Natural gas is often found in oil at high pressure in reservoir or as a gas cap over or above oil. As in case of oil, the major gas fields are also classified as Super Giants. India currently has 830 billion cubic metres of recoverable natural gas reserves. About 75% of this comes from Bombay High region and rest from Krishna, Cauvery and Assam basins.

(B) Metallic and Non-metallic Minerals

India is a poor country in mineral wealth. Except for iron ore and bauxite ore share world reserved of every other minerals is one percent or less. Will exhaust our reserves of all the important minerals and fuel except coal, iron ore, lime stone and bauxite in 25 to 30 years.

Use and Exploitation

Minerals are widely used in developed countries. The distribution of mineral resources on the earth are not equal. There are some countries have a large deposition of mineral wealth and others have no deposition. India is rich in mineral resources. The minerals have different quality. The use and exploitation affects our future generation. So these resources have some limits. the regular exploration of minerals depleting our available resources. Increasing human population growth and a large number of industries play a vital role in depleting it. Petroleum, tin, nickel, zinc and lead these minerals are hardly occur in our India. But bauxite, coal, mineral oil and iron found in plenty. Aluminium is extracted from bauxite. Copper deposits found in Bihar and Rajasthan. Reserve of uranium and thorium used in generation of atomic energy.

Increased population growth requires high consumption of minerals so it is necessary to conservation of mineral resources. Shortage of many minerals fullfilled by the search of new resources.

Environmental Effects of Extracting and Using Mineral Resources

Mining and processing have a significant effect on land, air, water and biological resources. Some effects of mining are given below:

(*i*) *Energy.* Energy leads to impacts such as acid rain and global warming.

(*ii*) *Impact on the Biological Environment.* Biological environment affected directly or indirectly. Directly like death of plants or animals owing to mining activity or contact with toxic soil or water from mines.

Alterations in groundwater or surface water availbility have indirect impacts like changes in total biomass, species diversity, nutrient cycling and ecosystem stability.

(*iii*) *Pollution.* Mining processes polluting our surroundings like atmosphere, surface waters and groundwater. Several elements like Cd, Co, Cu and Other leached from mining wastes and accumulates in water, soil or plants and produces different types of ill effects who drinking such contaminated water. Explosions, processing and transportation lead to the death of surrounding vegetation.

(*iv*) *Noise.* Blasting and transport are noise creating process and disturb local residents and our wildlife.

(*v*) *Dereliction of Land.* Large mining operations disturb the land by directly removing material in some areas and by dumping wastes in others. There can be a considerable loss of wildlife habitat. Mining activity also has a impact on land degradation.

(*vi*) *Supplies of Mineral Resources.* A large amount of minerals by extraction and processing to make products. Developed countries need to change from a high-waste and developing countries need to insure that they do not accept such an approach. Low waste process mainly depends on recycling, reusing and waste reduction.

(*vii*) Several physical changes in the land, water, soil and air are related to mining.

(*viii*) Our historical monuments also affected by mining processes.

(*ix*) Mining leads to deforestation, in this way flora and fauna also depleted.

(*x*) Daily life of human beings also affected by mining processes.

Case Studies. Aravallis cover about 10% of the geographical area in Rajasthan and are spread over 550 km. These mountains are rich in mineral wealth. There are about 9,891 mines in this region. The Hon'ble Supreme Court's decision on Nov. 20, 2002, led to closure of all these mines in Rajasthan. Rajsamand lake was totally dried up two years ago after a decade of marble mining in Rajnagar area. The lake never dried up completely in the past three hundred years. Though the blanket ban was lifted in December, 2002 stating that mining would be permitted in forest areas where prior permission had been obtained under the Forest and Environment Protection Act. But it was noted that during the seven days the mines were closed, the water level in the wells went up in the Rajsamand area.

FOOD RESOURCES

The main sources of human food are plants and animal. We consume almost all parts of one or the other plant in the form of cereals, pulses, vegetables, fruits and spices. A number of animal product such as milk, butter, egg and meat also supplement our food requirements. Since the major source of our food are plants, it becomes necessary to increase the production of crops to feed the ever increasing human population, the cropping pattern is determined mainly by natural factors like rainfall, climate and soil condition. The technological factors (*i.e.*, improved seed-fertilizer technology) plays a significant role in crop yield. Modern agriculture largely based upon technological factors like the use of improved seeds, chemical fertilizers, synthetic pesticides and extensive irrigation.

Problems

Though technological factors have proved a boon to our modern agriculture and have increased the crop yield manifold, they have created a number of problems.

(*i*) Intensive farming have reduced the fertility and productivity of the soil, and one can't think of crop yield without the use of chemical fertilizers.

(*ii*) Irrational use of chemical fertilizers to boost up the crop yield, have contaminated groundwater with nitrate. The presence of excess nitrate in drinking water is dangerous for human health and may be fatal for infants. Nitrate reacts with hemoglobin and impairs the oxygen transport by the blood. This condition is called **methaemoglobinemia** or **blue baby syndrome.**

(*iii*) The excessive use of pesticides and herbicides have depleted the population of ecologically important soil micro-organisms. Many of these chemical pesticides are non-degradable and enter the food chain, and become hazardous to human life.

(*iv*) Intensive irrigation (*i.e.*, paddy crop), is bringing the underground soluble salts to the solid surface and increasing the salinity of the soil. A large area of our fertile land have become saline and water logged in the recent years due to excessive irrigation.

CHANGES CAUSED BY AGRICULTURE AND OVERGRAZING

Food is a vital necessity of our life. There are many peoples in the world live on farming. But production, processing and distribution of food produces larger effects on the environment.

Changes Caused by Agriculture

Agriculture has two types of effects—primary and secondary. Primary effect where the agriculture takes place or on site effect. Next secondary effect away from the agricultural site known as off site effect. Finally effects are further divided into three groups—global, regional and local.

(*i*) *Global Effects*. In this type, effects due to climatic changes or extensive changes in chemicals cycles.

(*ii*) *Regional Effects*. Regional effects are developed due to combined effects of farming practices in the same wide area. Large scale pollution, deserification, deforestation, increase in sedimentation in major rivers and in the estuaries at the mouths of the rivers and changes in the chemical fertility of soils over large areas these different effects comes under the regional effects. In tropical waters, sediments entering the ocean can destroy coral reefs.

(*iii*) *Local Effects.* These effects occur near the site of farming. These effects like soil erosion, and increase in sedimentation, downstream in local rivers. Chemical fertilizers carried by sediments and produces eutrophication of local water bodies and toxins destroy local fisheries.

Changes Caused by Overgrazing

Fertility of soil and rainfall are two factors which affects carrying capacity of land. The land is quickly overgrazed when the rate of carrying capacity is exceeded.

The changes due to overgrazing are given below:

(*i*) Dominance of plant species that are relatively undesirable to the grazing animals.

(*ii*) Reduction in the growth of vegetation.

(*iii*) Diversity of plant species is decreases.

(*iv*) Increased soil erosion as the plant cover is decreases.

(*v*) Cattles damages land area due to trampling like paths made by cattle develop into gullies, which erode rapidly in the rain.

EFFECTS OF MODERN AGRICULTURE

Modern Agriculture have two types effects, one is positive and another negative. Agriculture sector use chemical fertilizers, which produces revolution in agriculture in the short-term, but its long-term effects extremely undesirable. According to environmentalists, such chemicals spread through the environment and pose a potential threat to all animals. Effects of modern agriculture are related to fertilizers, pesticide, waterlogging and salination.

1. **Fertilizer Problems.** Fertilizers are used for improving soil fertility. Fertilizers are two types—natural and chemicals. So the water sources get contaminated by leaching of excess fertilizers. These chemical compounds are not easily degraded. They either persists or accumulates and pass through the different trophic levels of the food chain. Chemical fertilizers may be serious source of pollution. Concentration of nitrates exceeds 25 mg/L, produces a serious health hazard disease called as "Blue Baby Syndrome" or methaemoglobinemia. In this way phosphates are also harmful for water bodies. If the level of phosphates increased, algal blooms developed in large numbers, which is responsible for 'eutrophication'.

2. **Pesticide Problem.** Chemical compounds that are used for the control of pests are called as pesticides. These are used for food production and killing insects, acting as vectors of many human diseases such as malaria, dengue, filaria and plague etc. Pesticides are further divided into six types: (*i*) Insecticides (*ii*) Weedicides (*iii*) Fungicides (*iv*) Rodenticides (*v*) Molluscicides (*vi*) Nematicides.

   Commonly used pesticides like—DDT, organophosphates, agricides and insecticides. D.D.T., aldrin and dieldrin are harmful to their persistent nature.

High concentration of DDT in human beings produces cerebral hemorrhage, liver damage, cancer, hypertension etc. Pesticides effects ecological balance by upsetting the pre-predator population. Human health also affected by short terms and long-term basis. In the short-term basis acute poisoning and illness due to high doses and accidental exposures. Next long-term basis include Parkinson's disease birth defects, growth of cancer and other chronic regenerative diseases. Long terms effects due to very low doses of different chemicals.

3. **Salination and Water Logging.** When the concentration of soluble salts increases, this is known as salination. High temperature, low rainfall and poor drainage are the main cause of salintion. Minerals in the soils and exposed rocks of the earth's crust produces soluble salts.

Good agriculture land turns saline every year; in Punjab alone. Salinity may be checked by improving the drainage and the salinated lands reclaimed by the simple process of leaching with excess of water. In this way salts are leached down to greater depths.

Water logging occurs due to the rise in water-table or seepage. The pores of soil are filled with water and produces oxygen deficiency. This may cause less transport of water and nutrients through the plant roots, reduction in root respiration and formation of toxic compounds in soils and plants. In this way suffocation produces in salinity.

Salinity. Salinity is mainly depends on concentration of soluble salts. It increases with the rate of evaporation. High temperature evaporates water and leaving behind the salt. Plants are mainly affected due to this and reduced their osmotic potential of the soil solution and the toxicity of specifications such as sodium, chloride and boron.

Measuring Salanity. The salanity of the soil is estimated by methods based on the ability of the salt in the soil solution extracted from a saturated sample of oil give an indication of the salt level.

Growth of Plants on Saline Soil. The soil solution consists of a large number of dissolved salt is brought into contact with a plant cell, water will pass by osmosis from the cell into the more concentrated soil solution. The cell then collapse. Sodium dominated soils produces O_2 deficiency due to breakdown of soil structure. In this way plants are affected.

Control. The retardation of evaporation is an important fact in the control of salinity. Low evaporation rate save moisture and also retard the translocation; upward of soluble salts into the root zone. For the dilution of salts requires light irrigation, which allow normal plant growth. Production of salt resistant crops are important site for management of salinity. Farm manure is very useful in such attempts.

Case Studies. Water-logging and salinity in Punjab, Haryana and Rajasthan. The first report of salt affected wasteland formation in connection with irrigation practices come from Haryana in 1958. It was reported that several village in area of Panipat, Rohtak and Delhi lying command area of western yamuna canal suffering from destructive saline efflorescence. The "Reh committee" in 1986 drew the attention of the government authorities on some vital points showing a close relationship between irrigation, drainage and spread of "Reh" and "Usar" soils.

Energy Resources. Energy can be defined as the capacity to do work. Work is the product of force. The agent used for 'pushing or pulling' an object is called force. Thus, energy is abstract concept. You cannot see it or feel it, even if you have to pay for it. It is an important factor in the development of an economy. The demand for energy has increased with the economic development of the world. Like other resources, energy resources are also non-renewable as well as renewable.

Growing Energy Needs. It is an important factor in the development of an economy. The demand for energy has increased with the economic development of the world. In India, the per capita energy consumption is only 221 kg of oil equivalent year, a very low rates compared with other developed countries.

RENEWABLE AND NON-RENEWABLE ENERGY RESOURCES

Energy sources can be grouped into two important groups. These are Renewable and Non-Renewable. Their categories are as follows:

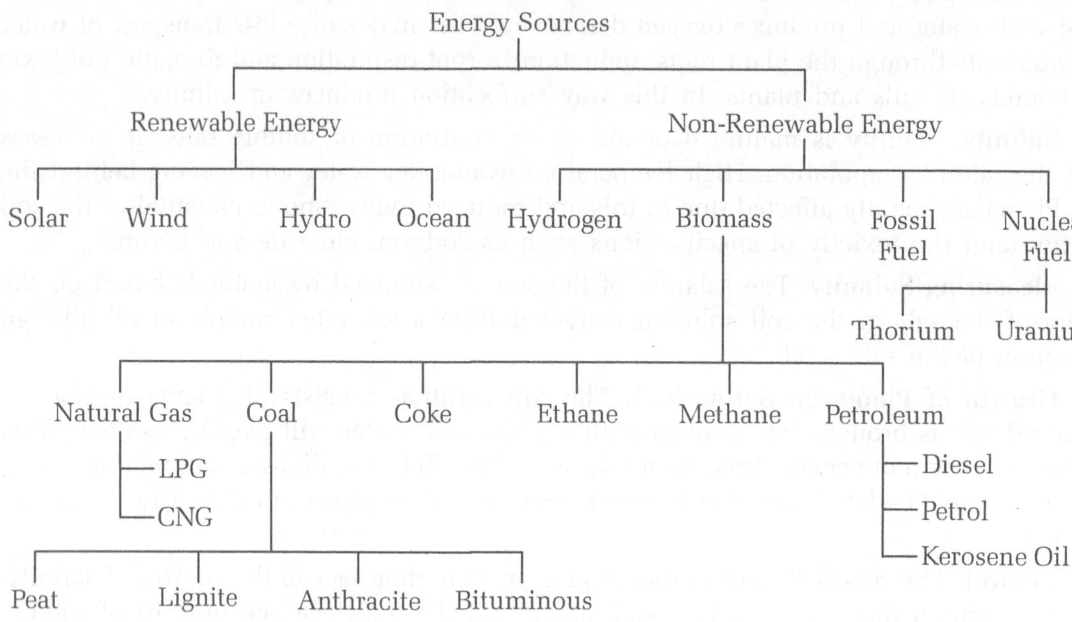

(I) Non-Renewable Energy Resources

These include various fossil fuels and nuclear energy. The fossil fuels include coal, petroleum products and natural gas. The nuclear energy is mainly obtained from the nuclear fission of the uranium. The world's reserve of fossil fuels and uranium are limited and eventually be exhausted. With vast expansion of industry and agriculture sectors, the available fossil fuels and uranium, began to fall in their supply. Besides, the burning of fossil fuels is causing a number of negative environmental consequences such as global warming, acid rain, air pollution and oil spills during transportation. It has, therefore, become necessary to minimise the use of fossil fuels and replace them with ecofriendly renewable resources.

(II) Renewable Energy Resources

These resources are regenerated by natural processes so that, they can be used indefinitely. These include solar energy, hydropower, wind energy, geothermal energy, ocean waves and tidal energy. The renewable energy resources generally cause much less environmental impact than fossil fuels or nuclear energy. Presently the generation of renewable energy is often expensive than the energy produced by fossil fuels or nuclear energy. However, with the advancement of technology, the cost of renewable energy production is expected to decrease.

Use of Alternate Energy Resources. Non-renewable energy system based on fossil fuel will be replaced and sustainable energy system based on renewable energy resources come to force.

1. **Solar Energy.** Solar energy renewed from incoming sunlight. When sun rays strikes any solid or liquid, it gets absorbed and turned into heat energy. Solar energy used in many fields like, solar cooking, solar greenhouses, solar water heating, solar furnaces, solar pumping, solar heating of building, solar electric power generation, solar distillation, etc.
2. **Wind Energy.** Wind energy is known as non-pollution energy and can be harnessed by using wind mills. A water pumping windmill consists of a rotar, tower pump rod, transmission system and pump. The energy of the wind is intercepted by the rotar and rotar moves is in circular motion. The rotar is mounted on a shaft. The rotatory motion of the shaft makes the pump rod move in an up and down motion causing water from well to get lifted to ground level.
3. **Geothermal Energy.** Geothermal power is obtained by means of heat engines from heated water and steam tapped through drilling holes at selected places. This heat is utilized for the power generation. Geothermal energy used for greenhouse heating and building.
4. **Biogas Energy.** Biomass leads to a separate category of energy derived from organic wastes from agricultural wastes such as animal dung, bagasse etc. Biogas generated from anaerbic decomposition of organic wastes, which contain methane and carbon dioxide. Biogas can be used for cooking purposes, lighting, space heating and in biogas plants.

Case Studies. The first nuclear power station in India was constructed at Tarapur in Thana District near Mumbai, which is generating electricity since October 1969. At present, the capacity of nuclear power plants in the country is 1940 MWe (Mega Watt electrical) accounting for about 2% of the total power generating capacity in the country. The operating atomic power stations are Tarapur Rawathbatta, Kalapakkam, Narora and Kakrapora. The department of Atomic Energy set up in 1954 is the executive agency for all activities related to nuclear energy.

LAND RESOURCES

Land forms about 20% of earth's surface about 13,393 million hectares. About 36.6% of the land area is covered by houses, factories, roads, railways, deserts, mountains, rocks, glaciers and polar ice marshes. About 30% of land is occupied by forests. About 22% of land is occupied by meadows and pastures. Only 11% of land is fit for ploughing. The surface layer of land is called soil. About four-fifth of land consists of soil.

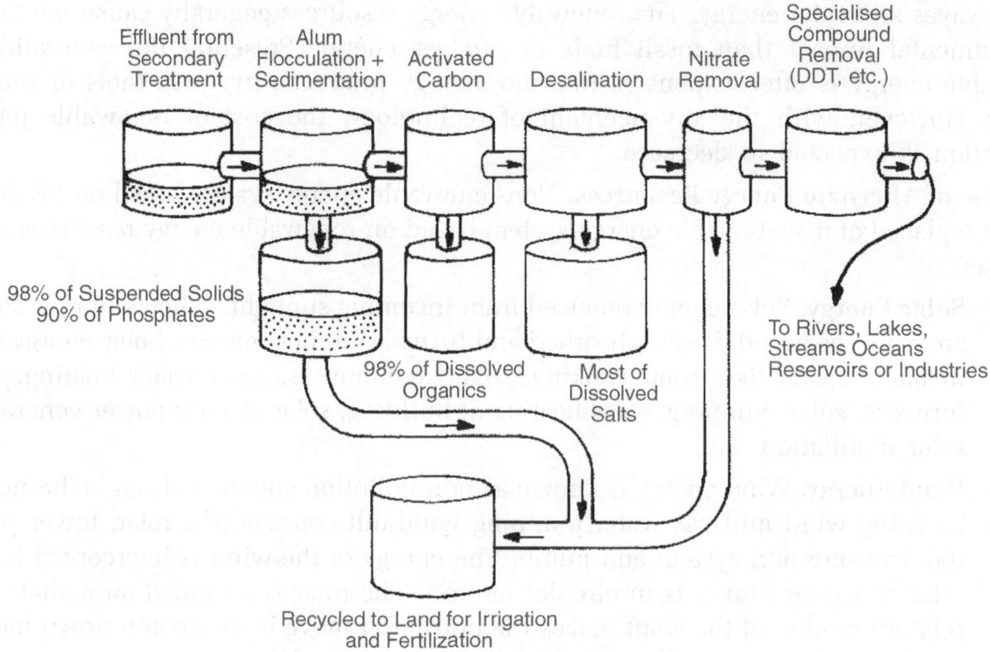

Fig. 2.3. *Diagrammatic Representation of Tertiary Treatment in an Efluent Treatment Plant.*

Soil is the upper weathered layer of earth's crust which supports plant life and contains many living plant organisms and their dead parts. The study of soil is called **pedology** (Gk *pedion* = ground, *logos* = discourse). It deals with origin, formation and geographic distribution of the soil.

Soil is a most important resource and it takes decades or even centuries for the development of soil horizons having physiochemical properties. Soil is formed from two processes—weathering (breaking down of rocks into small practices) and *pedogenesis* (maturation of soil through development of humus). Weathering is caused by several forces—tempertature changes, alternate drying-wetting, mechanical action of water and sand, chemical reactions, activity of plant roots, lichens and microbes.

Soil Erosion. The removal of top soil from its place by various physical agencies like wind, water and ice etc., is called soil erosion. The presence of plant cover significantly reduces soil erosion. Human activities accelerate soil erosion by removing natural plant cover.

Land Degradation. Land degradation known as loss of fertility or productive capacity of the soil. The factors which are responsible for land degradation are given below.

1. Soil erosion, 2. Desertification, 3. Soil pollution, 4. Shifting cultivation, 5. Salination and waterlogging, 6. Urbanisation.

1. **Soil Erosion.** Soil erosion is a natural process. The loss of top fertile soil by the action of water, wind or by the activities of man is known as soil erosion.

2. **Desertification.** Accelerated erosion by wind and water, woodland destruction salinization, waterlogging and excessive grazing by livestock's are the main factors for desertification. Desertification is a serious and major problem in dry lands.
3. **Soil Pollution.** Soil pollutants reduces the productivity of soil. This is known as **soil pollution.** Excessive fertilizers, insecticides, pesticides, etc., have created the problems of soil pollution.
4. **Shifting Cultivation.** Destruction of forests in many hilly areas of India is a main reason for the soil erosion. This is known as 'Jhuming'.
5. **Urbanisation.** Developmental activities such as human settlement, dams, roads, airports, industries etc., are responsible for the depletion of productive area.

Control of Land Degradation: Land is an important non-renewable resource. Land management is necessary because the process of soil formation is very slow.

Following measures can be taken for land planning.

1. **Planting Wind Breaks and Shelter Belts.** Wind is the main factor for soil erosion. So wind breaks check the movement of wind. Rows of big trees also check the velocity of wind.
2. **Reforestation and Afforestation.** Reforestation means replanting forests where they were destroy overgrazing, forest fires etc. This is helps in checking soil erosion, waterlogging and floods. Afforestation means growing forests where were no forests before.
3. **Agricultural Practices**
   (*i*) *Crop Rotation.* Sowing of different crops in the same field in a regular sequence for a number of years decreases soil erosion and preserves the productivity of land.
   (*ii*) *Terracing.* A big sloping drainage area is divided into a no. of small separate flat fields known as **terraces.** This helps in the slow down the speed of run-off water.
   (*iii*) *Regular Cultivation.* Regular cultivation is necessary for checking erosion. In other way regular cultivation maximise the rate of water infiltration into the soil.

REASONS FOR LANDSLIDES

Human activities such as construction of road and railway, canal, dams, reservoir and mining operation in hilly areas have affected the stability of hill slopes and damaged the vegetation cover. These activities upset the balance of nature, making such areas vulnerable to landslides. Such activities also responsible in increasing the turbidity of various nearby stream, thereby reduced productivity.

Soil Erosion

Soil erosion results in the loss of the soil fertility. This is natural a process. Soil loss is maximum in region with high population densities. Soil erosion mainly of two types:

1. **Geologic Erosion.** It occurs in the presence of natural conations. It is very slow process.
2. **Accelerated Erosion.** This is the most serious type of loss and caused by an interference of man and other animals.

The factors which influencing the soil erosion are given below:

1. **Nature of the Soil.** Nature of soil affected the soil erosion. When soil exposed same type of rainfall and then light, open soils lose more silt than heavier loam's which swell up by wetting.
2. **Vegetation Cover.** Vegetation play a vital role in management of soil. Roots of the plants binds up the soil and restrict the soil erosion. On the other hand bare land where rainfall causes soil erosion due to loose soil easily carried away by rain water.
3. **Distribution of Rainfall.** Unequal distribution of rainfall harmful for soil. Because heavy rainfall produces excess of water and soil is unable to absorb this excess water. In this way this water removes the layers of soil resulting in soil erosion.
4. **Slope of the Ground.** Steep slopes are the main cause for soil erosion. Because the infiltration rate decreases and the run-off rate increases resulting in more soil erosion.
5. **Soil Mismanagement.** Soil erosion produces due to overgrazing, forest-fires, surface drainage, wrong methods of cultivation etc. Soil erosion mainly effected by uncontrolled grazing by animals produces a breaking in the vegetation cover. Continuous cultivation of same crop also adds to soil loss. Soil is washed or blown away faster than it can possible be regenerated and so there is not loss of soil resulting in changing of productive soil into desert like condition.

Millions of tons of top soil is washed away every year from our crop fields and is transferred to sea. Erosion causes significant loss of soil fertility by transporting organic matter and nutrients of the soil. The eroded soil, generally gets into streams, rivers and lakes in the form of sediments. It not only spoils the quality of water but also harms the habitats of aquatic organisms.

DEPLETION OF SOIL FERTILITY

The growing population and advancement of civilization have forced human beings to clear large tracts of natural vegetation of develop agricultural lands residential areas and industrial establishments. When natural vegetation is removed, not only the nutrients stored in vegetation are removed, the organic matter and nutrients accumulated in the soil are also lost. From agricultural systems nutrients are lost through crop harvest. As a result, the agricultural soil in availably loses its fertility over a period of time.

Water logging and application of non-degradable chemicals also leads to the loss of soil fertility.

Soil Conservation

Soil conservation refers to the measures and practices, which protect the soil against loss and help in maintaining its fertility with a view establish sustainable agriculture. The art of soil conservation is based on following basic principles :

(*i*) Protection of soil from impact of rain drops.

(*ii*) To slow down the water from concentrating and moving down the slope in a narrow path.

(*iii*) To slow down the water movement, when it flows along the slope.

(*iv*) To encourage more water to enter the soil.

(*v*) To increase the size of soil particle.

(*vi*) Reduction in the wind velocity near the ground by growing vegetation cover and ridging the land, and

(*vii*) To grow strips of stubble or other vegetation cover, which might catch and hold the moving particles of soil.

Keeping in view the above said principles, ecologists have devised several methods, to prevent the loss of the soil due to its erosion.

These methods include the following practices:

1. **Conservational Tillage.** The incorporation of residues from previous crops into the soil by ploughing is called Conservational tillage. It improves soil permeability and increases the organic matter, which in turn improve soil moisture and nutrients.
2. **Organic Farming.** Organic farming involves practices which provided increased organic input to the soil. Application of biofertilizers is an important practice of organic farming. The organic farming has long terms beneficial effects on soil fertility.
3. **Crop Reproduction.** It is a practice of growing different crops (generally legumes after cereal crops) in successive years on the same piece of land. It decreases soil loss and prevents depletion of nutrients from the soil.
4. **Contour Ploughing.** It is an old method useful in areas with low rainfall in the preparation of the field with alternate furrows and ridges. Water is caught and held in furrows and stored, which reduces run-off and erosion.
5. **Mulching.** Soil is allowed to remain untilled and is covered with crop residue and other form of plant litter. It retain soil moisture, decreases run-off and increases organic matter in the soil.
6. **Strips Cropping.** It involves planting of crop in rows or strips to check flow of water. It may be contour strip cropping (strip planted along contour at 90 degree to the direction of slope), field strip cropping (strip planted parallel to each other), or wind strip cropping (strip planted in straight parallel rows at 90 degree to the direction of prevailing wind).
7. **Terrace Farming.** A slope is divided into a number of small flat fields called terraces. The terraces slow down the velocity of run-off and allow the water to move to the sides of the fields, where it flows away without eroding the soil.

8. **Agrostological Methods.** In agrostological methods grasses are grown to check soil erosion. The grasses act as soil binders or stabilizers. The grasses are grown either in rotation or alongwith agricultural crops (lay farming) or they are grown on such lands, where major portion of the top soil has been eroded (retiring lands to grasses). The grasses prevent erosion and also improve soil permeability.
9. **Afforestation.** Trees or wind breaks are planted in deserts which check the velocity of wind. Wind breaks are planted across the area of 90 degree to the prevailing wind.

CONSERVATION OF EROSION-AFFECTED SOILS

The conservation of erosion affected soils involve following two steps:

(*i*) *Soil Stabilization.* Soil stabilization is achieved by seeding the bare ground with plants that can survive harsh conditions, *e.g.*, growing drought resistant grasses (lay farming). The growing grasses eventually establish vegetation cover on the soil, preventing further erosion.

(*ii*) *Restoration of Soil Fertility.* The soil fertility can be restored by increasing the amount of detritus to the soil. It increases nutrients and organic matter in the soil and also improve moisture level of the soil. The restoration of soil fertility to its original level is slow process.

DESERTIFICATIONS

Deserts have a combined effect on salinization water logging, wind erosion and overgrazing in dry land fields. The famous and well known **Thar Desert** in India developed by destroying thousands of hectares of productive land. Due to salinization the high concentration of salt produces impairment in plant development. Another factor is excessive grazing by animals resulting desertification. In the Western Rajasthan the hot arid regions are facing a major problem of desertification due to high speed of winds, salinization, overgrazing, waterlogging etc.

Deserts can be checked by Afforestation. Heavy herds of goats and other animals destroy the saplings planted. When erosion occur due to loss of top fertile soil produces desertification. Shifiting of sand dunes from one place to another, the conversion of fertile land into deserts occur each year. Many deserts are formed by the result of human activities.

Role of an Individual in Conservation of Natural Resources

Different types of natural resources are necessary for well being and development of a nation. However, over exploitation of these natural resources resulting fast depletion of these resources. In this way conservation of these natural resources are necessary. Natural resources gives us a large number of vital products, so it is our duty to conserve it. Role of individual in conservation of natural resources are given below.

A. Water Conservation

1. Check for water leakage in pipes and repair them quickly.
2. Another important way is to prevent wastage of rainwater. Rain replenishes water in tanks, lakes, pond and reservoirs, it runs-off into rivers and streams and seeps into the ground and is stored as groundwater.
3. Build rainwater harvesting system.
4. Stop wastes of water when water taps running.

B. Mineral Conservation

1. Modern technologies for recycling of metals.
2. Substitutes of fossil fuels need to be found.
3. Alloys are developed for reducing the demand of pure metals.

C. Energy Conservation

1. Heaters and cooling equipment should be correctly sized positioned.
2. High efficiency lighting system should be installed and kept clean.
3. Each energy operation should be examined critically and systematically in isolation and in relation to all other events occurring within system boundary.

D. Soil Conservation

1. Soil conservation through plant vegetation cover.
2. Growth of grasses play a vital role to check soil erosion.
3. Sprinkling irrigation protect our soil.
4. Crop rotation decreases soil loss and secure the productivity of land.
5. Do not uproot the trees near your house.
6. Over irrigation produces water-logging problem.

Equitable Use of Resources for Sustainable Lifestyles

Concept of sustainable development could be defined as the management and conservation of natural resources. Technological and institutional change in such as manner a to ensure both development and the continued satisfaction of human needs for the present as well as future.

Development should be planned and managed in an integrative and adaptive manner, and should operate from as solid and up to date an information base as possible. Sustainable development meets the needs of the present without compromising the ability of future generation to meet their own needs.

CONCEPT OF SUSTAINABILITY

The concept of sustainability reveals balance of economic, social and ecological goals. Other concept of sustainability around earth's carrying capacity and based on natural resources as ozone layer, fertile soil etc. Sustainable development also related to concepts of carrying capacity and green accounting, which are given below:

1. **Carrying Capacity.** Carrying capacity of a region may be defined as number of individuals of a species that it can sustain. Carrying capacity has two types, supporting and assimilative capacity. The supporting carrying capacity related to stock of available resources and assimilative capacity to pollution load.
2. **Green Accounting.** Green accounting widely known in developed and developing countries. This is mainly related to use of natural resources for improving the quality of environment. It making the task easier for planners and policy makers to formulate further programmes of development.

The concept of sustainable development should include:

1. Resource conservation.
2. Recycling and resue of material waste.
3. Management of renewable resources.

EXERCISES

1. What is meant by Natural Resources ? Write its main types ?
2. Give a detailed account of Forest Resources ?
3. Write a detailed note on Water Resources ?
4. Write a detailed note on Mineral Resources ?
5. Write short note on Marine Mineral Resources ?
6. Write a detailed note on Food Resources ?
7. Write a detailed note on Energy Resources ?
8. Discuss the Land Resources ?

❑❑❑

Unit III ECOSYSTEM

MEANING OF ECOSYSTEM

The terms ecosystem was proposed by **A.G. Tansley** in 1935. Who defined it as *'the system resulting from the integration of all the living and non-living factors of the environment'*. Thus, he regarded the ecosystem is including not only the organism-complex but also the whole complex of physical factors forming the environment. However, the terms ecosystem is most preferred, where *'eco'* implies the environment and *'system'* implies an interacting, interdependent complex.

Odum (1963) defined it as *"An Ecosystem is a self-sustaining energy driven system"*.

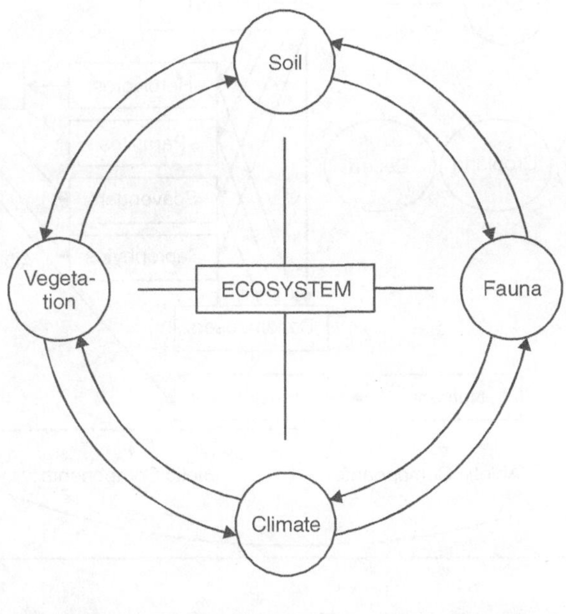

Fig. 3.1. *Relationship between Soil, Vegetation, Climate and Fauna*

CONCEPT OF ECOSYSTEM

An ecosystem is the whole biotic community in a given area plus its abiotic environment. Ecosystem ecology emphasises the movements of energy and nutrients (chemicals elements) among the biotic components of ecosystems. Nutrients are cycled within ecosystems to a much greater extent.

All living organisms and their environment are mutually reactive affecting, each other in various ways. Environment, which is actually a complex of several inter-related factors and is much dynamic the species puts each effort to maintain its uniformity in structure, function, reproduction and growth by preservation of its genetic pool. The rate of production *i.e.,* amount of organic matter accumulated in the living component of an ecosystem in unit time is referred to as the productivity of the ecosystem. In any ecosystem, various living organisms are arranged in a definite sequence according to their food habits. With the new concept, both biotic and abiotic components are considered, the basic structural and functional units of nature are ecosystem. These exist varying degrees of +– or even neutral interactions among organisms, at both, inter and intraspecific levels, which determine along with abiotic parameters. But energy play a very important role in ecosystem, energy is the driving force of this system.

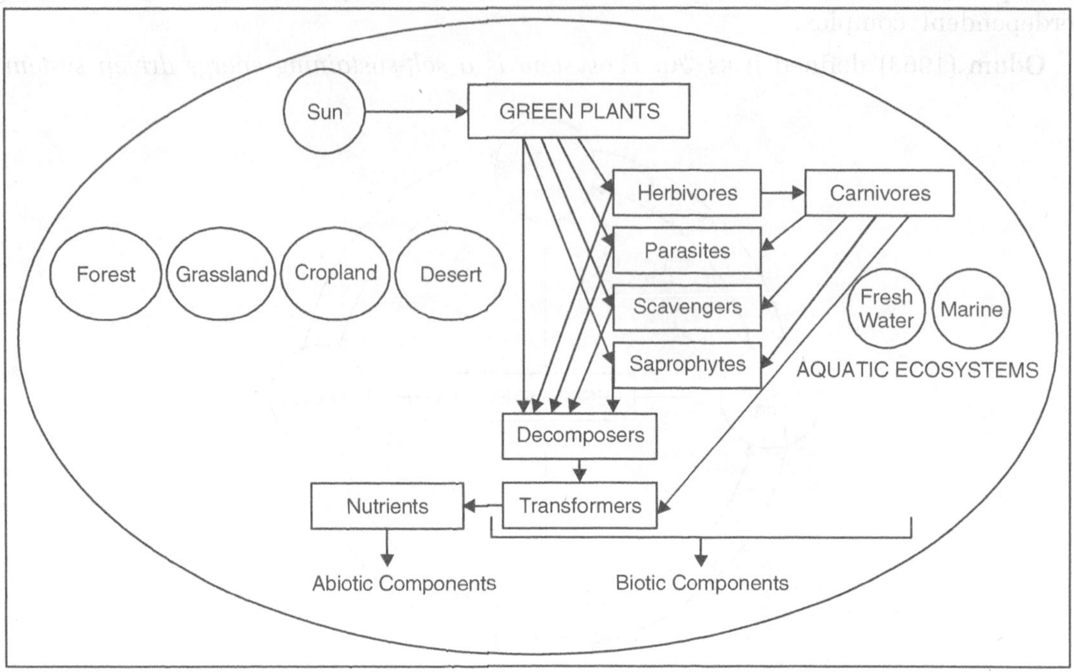

Fig. 3.2

Ecosystem summarised in the following manner:

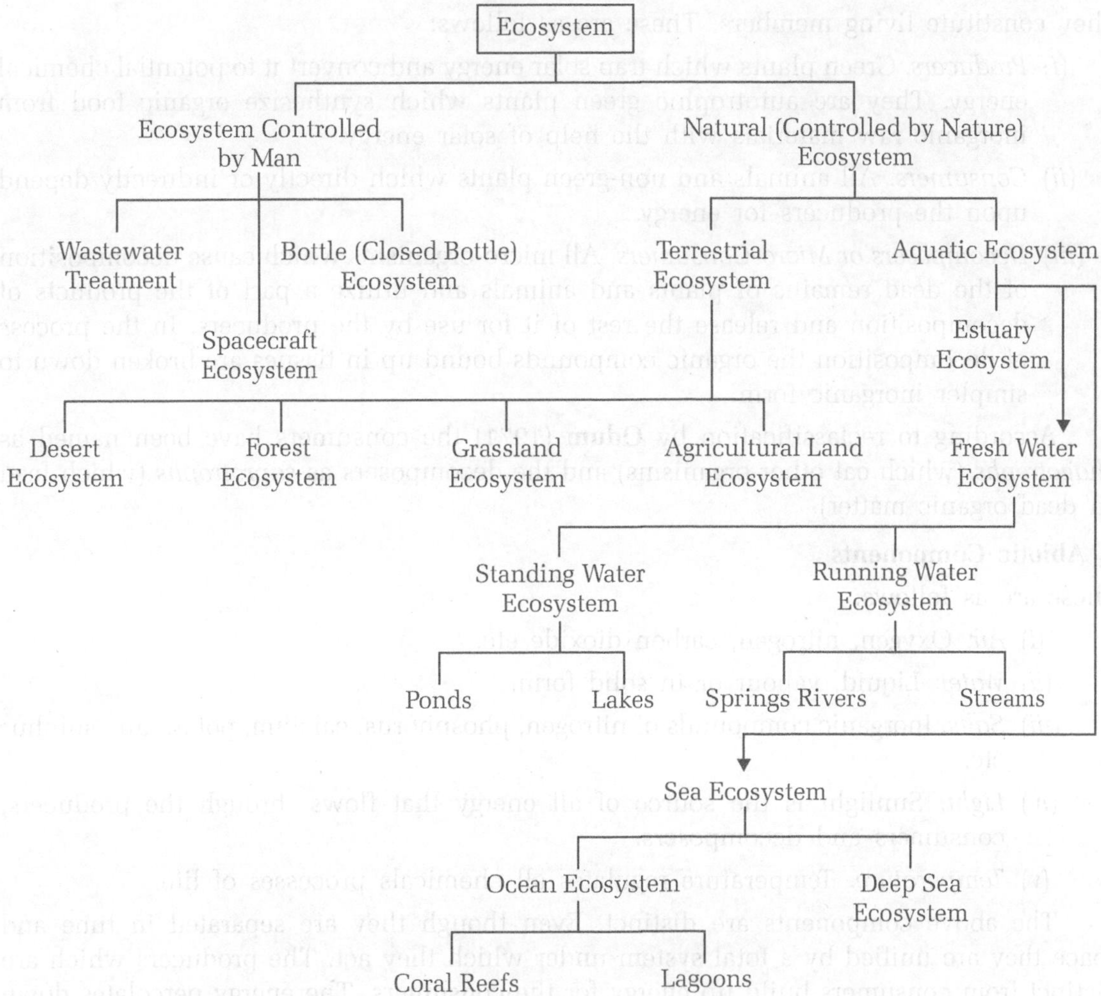

STRUCTURE AND FUNCTION OF AN ECOSYSTEM

Components such as light, air, water, inorganic elements constitute the abiotic components. The biotic components which constitute the living organisms depend upon these abiotic components. The living organisms influence the inorganic components by their organic compounds and check deterioration of the ecosystem. In this way, in a healthy habitat the biotic and abiotic components interact in such a manner that a perfect balance is maintained. This phenomenon is called homeostasis or balance of nature. It fluctuates within certain limits as it is not static in nature. Homeostasis is maintained through a number of controls (= limitations = cybernetes) such as: (*a*) Carrying capacity of the environment, (*b*) Capacity of the ecosystem for recycling of wastes. (*c*) Self-regulation (*d*) Feedback system.

1. Biotic Components

They constitute living members. These are as follows:

(*i*) *Producers.* Green plants which trap solar energy and convert it to potential chemical energy. They are autotrophic green plants which synthesize organic food from inorganic raw materials with the help of solar energy.

(*ii*) *Consumers.* All animals and non-green plants which directly or indirectly depend upon the producers for energy.

(*iii*) *Decomposers or Micro-Consumers.* All micro-organisms which cause decomposition of the dead remains of plants and animals and utilize a part of the products of decomposition and release the rest of it for use by the producers. In the process of decomposition the organic compounds bound up in tissues are broken down to simpler inorganic form.

According to reclassification by **Odum** (1971) the consumers have been named as *phagotrophs* (which eat other organisms) and the decomposers as *saprotrophs* (which feed on dead organic matter).

2. Abiotic Components

These are as follows:

(*i*) *Air.* Oxygen, nitrogen, carbon dioxide etc.

(*ii*) *Water.* Liquid, vapour or in solid form.

(*iii*) *Salts.* Inorganic compounds of nitrogen, phosphorus, calcium, potassium, sulphur etc.

(*iv*) *Light.* Sunlight is the source of all energy that flows through the producers, consumers and decomposers.

(*v*) *Temperature.* Temperature regulates all chemicals processes of life.

The above components are distinct. Even though they are separated in time and space they are unified by a total system under which they act. The producers which are distinct from consumers build up energy for the consumers. The energy percolates down to different levels of consumers through food chain. The decomposers ultimately break down the complex organic compounds into simpler inorganic forms so that the producers can use them again.

In conclusion, the following facts about the ecosystem must be remembered:

(*i*) Any substance which helps in the sustenance of life is a components of the ecosystem.

(*ii*) Organisms are the components of the ecosystem. They derive substances from inorganic materials.

(*iii*) Life signifies transformation of energy.

(*iv*) Sun is the source of energy which cannot be created, cannot be destroyed. It is only transformed from one form to another form in an ecosystem.

Structure of Ecosystem

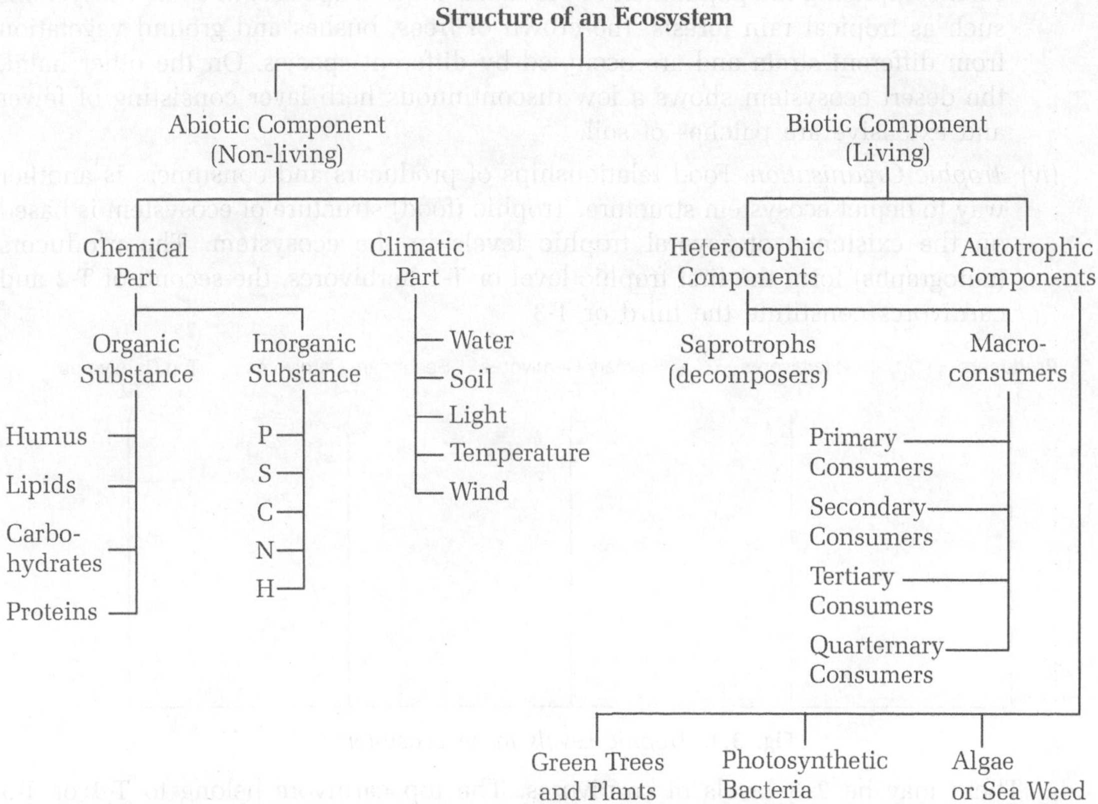

The structure of an ecosystem is characterized by the physical organisation of biotic and abiotic components. The major structural features of ecosystem are: species composition, stratification, trophic organisation and nutrients.

(*i*) *Species Composition.* Each ecosystem has its own type of species composition. Different ecosystem have different species composition. A great variety of species is found in forest ecosystem, whereas a few species occurs in a desert ecosystem.

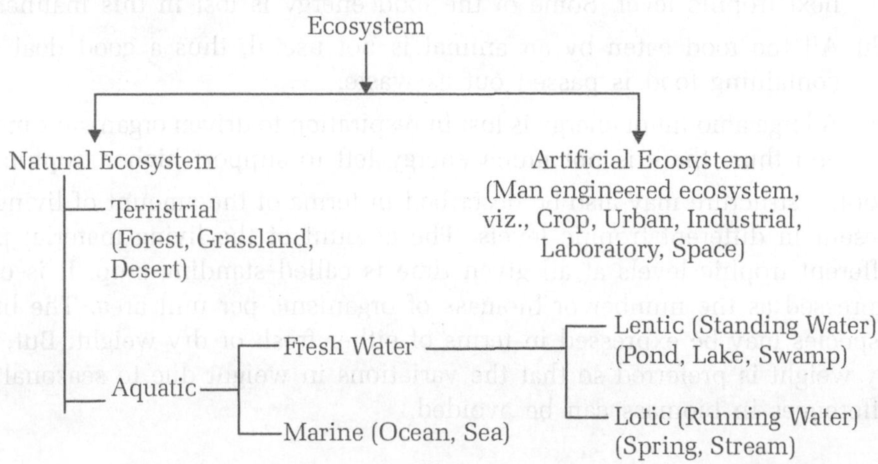

(*iii*) *Stratification.* The organisms in each ecosystem from one or more layers or strata, each comprising the population of particular kind so species. In some ecosystems such as tropical rain forests, the crown of trees, bushes and ground vegetation from different strata and are occupied by different species. On the other hand, the desert ecosystem shows a low discontinuous herb layer consisting of fewer and extensive are patches of soil.

(*iv*) *Trophic Organisation.* Food relationships of producers and consumers is another way to depict ecosystem structure. Trophic (food) structure of ecosystem is based on the existence of several trophic levels in the ecosystem. The producers (autographs) form the first trophic level or T-1 herbivores, the second of T-2 and carnivores constitute the third or T-3.

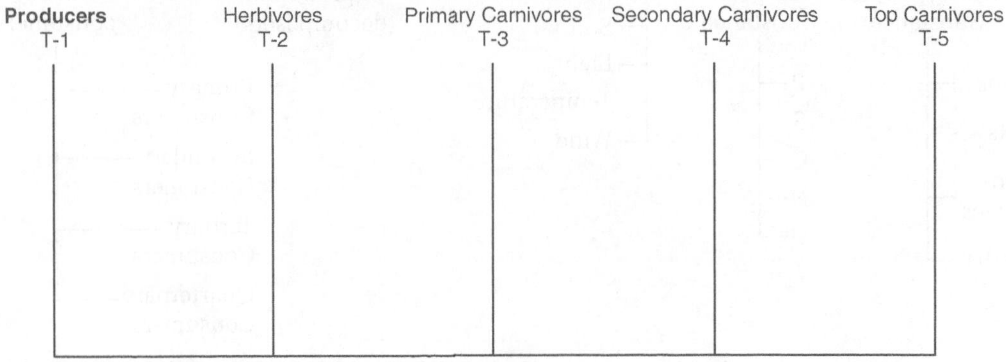

Fig. 3.3. *Trophic Levels in an Ecosystem.*

There may be 2–3 levels of carnivores. The top carnivore belong to T-4 or T-5 trophic levels. Decomposers form the ultimate or detritus trophic level. Parasites feed upon the organisms of all trophic levels. Therefore, they do not have a fixed trophic level. Similarly there is no fixed trophic level of omnivorous organisms like man.

In an ecosystem there can be only 4–5 successive trophic levels, because:

(*a*) All the food available at one trophic level is not eaten by the animals of the next trophic level. Some of the food energy is lost in this manner.

(*b*) All the food eaten by an animal is not useful, thus a good deal of energy containing food is passed out as waste.

(*c*) A large amount of energy is lost in respiration to driven organism's metabolism, and thus, there is not much energy left to support higher trophic levels.

Trophic structure may also be described in terms of the amount of living material present in different trophic levels. The amount of the living material present in different trophic levels at all given time is called standing crop. It is commonly expressed as the number or biomass of organisms per unit area. The biomass of a species may be expressed in terms of either fresh or dry weight. But, generally dry weight is preferred so that the variations in weight due to seasonal moisture differences in biomass can be avoided.

(*iv*) *Nutrients.* In each ecosystem, the nutrients necessary for the growth of living organisms are accumulated in the biomass and the abiotic components like soil. The amount of nutrients such as nitrogen, phosphorus and calcium present in the soil at any given time is termed as standing state. Different ecosystems have different standing state of nutrients. The standing state of nutrients may vary at different time even in the same ecosystem.

FUNCTIONS OF ECOSYSTEM

Ecosystems have some functional attributes which keep the component parts running together. Each ecosystem possesses a natural tendency to persist. Various functions performed by the structural components of the ecosystem ensure persistence of the system. For instance, green leaves prepare food and roots absorb nutrients from the soil. Herbivores feed on part of the plant production, and in turn serve as food for carnivores. Decomposers carry out the function of breaking down complex organic materials into simpler inorganic products which can be used by the producers. All these functions in the ecosystem occur through delicately balanced and controlled process. For example, food production in plants is carried out by the process of photosynthesis, nutrients contained in the organic matter are released through the process of decomposition. A knowledge of the rates of different processes at which they occur in the ecosystem is necessary to understand the interrelations, structure and functions of the ecosystem. The major functional attributes of the ecosystem are: (*i*) Productivity and decomposition (*ii*) Energy flow (*iii*) Nutrient cycling (*iv*) Development and stabilization.

Producers. Producers are those organisms that derive energy from either sunlight or inorganic compounds and makes organic compounds from inorganic compounds. Example—many primary producers like green plants, algae, photosynthetic bacteria obtain energy from sunlight for making organic compounds oxidizing bacterial use energy stored in the chemical bonds of inorganic compounds for making organic compounds. As the green plants manufacture their own food, they are called as autographs.

The chemicals energy stored by the primary producers is utilized partly by themselves for their growth and survival while the remaining is stored in the plant parts for their future use.

Consumers. Consumer are those organisms which use organic compounds for their food and energy. They cannot make organic compound from inorganic compounds as they lack chlorophyll but they can transform on form of organic compounds into other forms of organic compounds.

Example. Animal, protozoa and many bacteria. They depends on the producers for their food and are called as heterotrophy.

The Consumers are of many types:

(*i*) *Primary Consumers.* These are the animals that feed on primary producer and are called as herbivores, example: Deer, goat etc.

(*ii*) *Secondary Consumers.* These are the animals that feed on the herbivores example : Hawk, lizard etc.

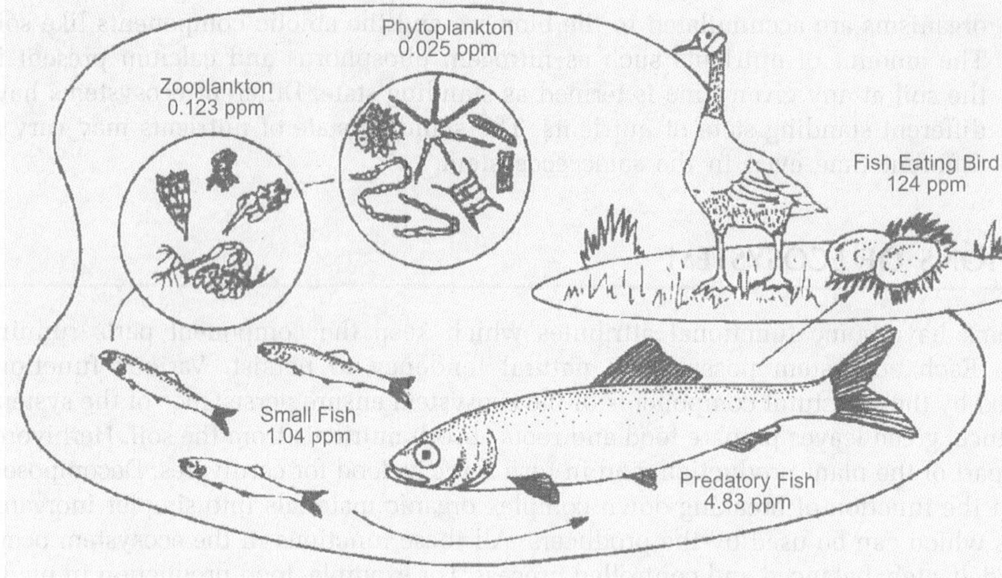

Fig. 3.4. *Process of biological magnification ; DDT concentrations increase in organisms along the food chain.*

(*iii*) *Tertiary Consumers.* These are the large carnivores that feed on the secondary consumer, example: Snake, Wolves etc.

(*iv*) *Quaternary Consumers.* These are the larger or top carnivores that feed on the territory consumers, example: Lions, Tigers etc.

Decomposers. Decomposers are also heterotropic organisms but depend upon dead organic matter for their food. They are chiefly micro-organisms like bacteria, actinomycetes and fungi.

The breakdown complex organic matter like bacteria, cellulose, hemicellolose etc. found in the plant and animals body, and ultimately release simple substances used by the producers.

ENERGY FLOW IN ECOSYSTEM

Ecosystem energetic in ecosystem energetic we study:

1. Quantity of solar energy reaching an ecosystem.
2. Quantity of energy used by green plants for photosynthesis and
3. The quantity and path of energy flow from producers to consumers.

About 34% of the sunlight reaching the earth's atmosphere is reflected back into its atmosphere, 10% is held by ozone layer, water vapour and other atmospheric gases. The rest, 56% reaches the earth's surface.

Energy Flow in Ecosystem. The behaviour of energy in ecosystem can be termed energy flow due to unidirectional flow of energy.

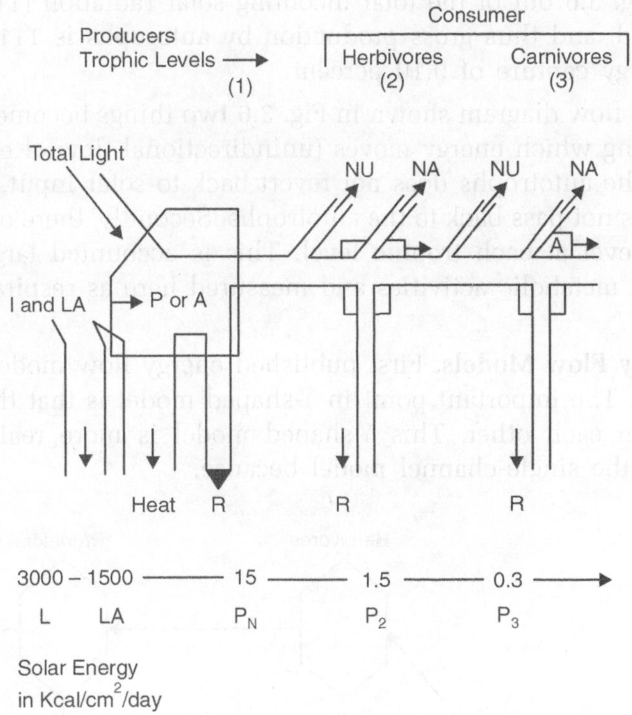

Fig. 3.5. *Energy Flow in an Ecosystem.*

Single-channel Energy Models. The Principle of food chains and the working of the two laws of thermodynamics can be better made clear by means of energy flow diagram shown in Fig. 3.5.

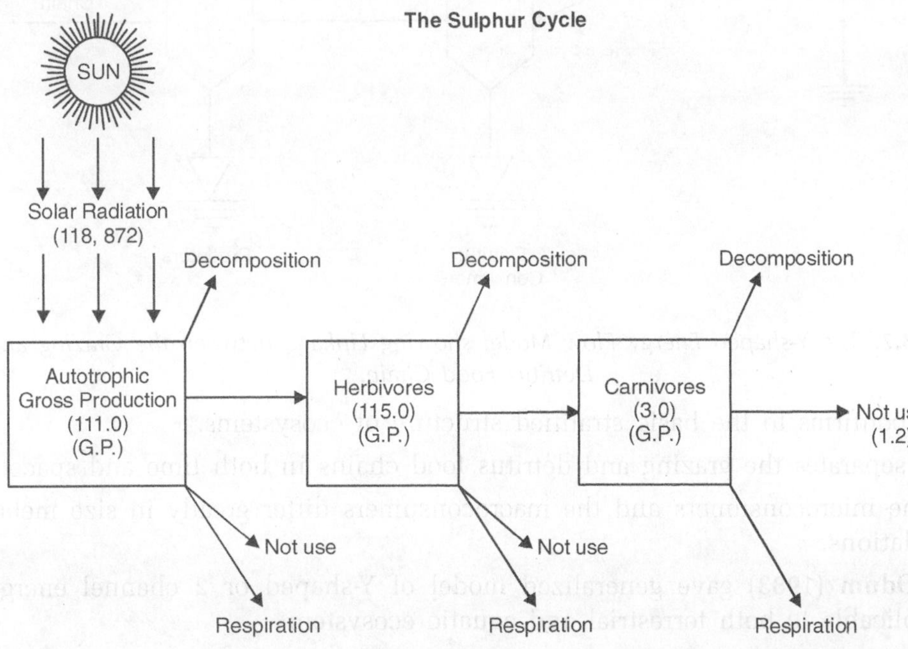

Fig. 3.6. *Energy Flow Diagram for Fresh Water in g cal/cm²/yr.*

As shown in Fig. 3.6 out of the total incoming solar radiation (118, 872 g cal/cm 2/ yr.) remain unutilized, and thus gross production by autorophs is 111 g cal/cm2/yr with an efficiency of energy capture of 0.10 percent.

From the energy flow diagram shown in Fig. 3.6 two things become clear, firstly, there is one-way street along which energy moves (unindirectional flow of energy). The energy that is captured by the autotrophs does not revert back to solar input, that which passes to the herbivores does not pass back to the autotrophs. Secondly, there occurs a progressive decrease in energy level at each trophic level. This is accounted largely by the energy dissipated as heat in metabolic activities and measured here as respiration coupled with unutilized energy.

Y-shaped Energy Flow Models. First published energy flow models as pioneered by **H. T. Odum** in 1956. The important point in Y-shaped model is that the two food chains are not isolated from each other. This Y-shaped model is more realistic and practical working model that the single-channel model because,

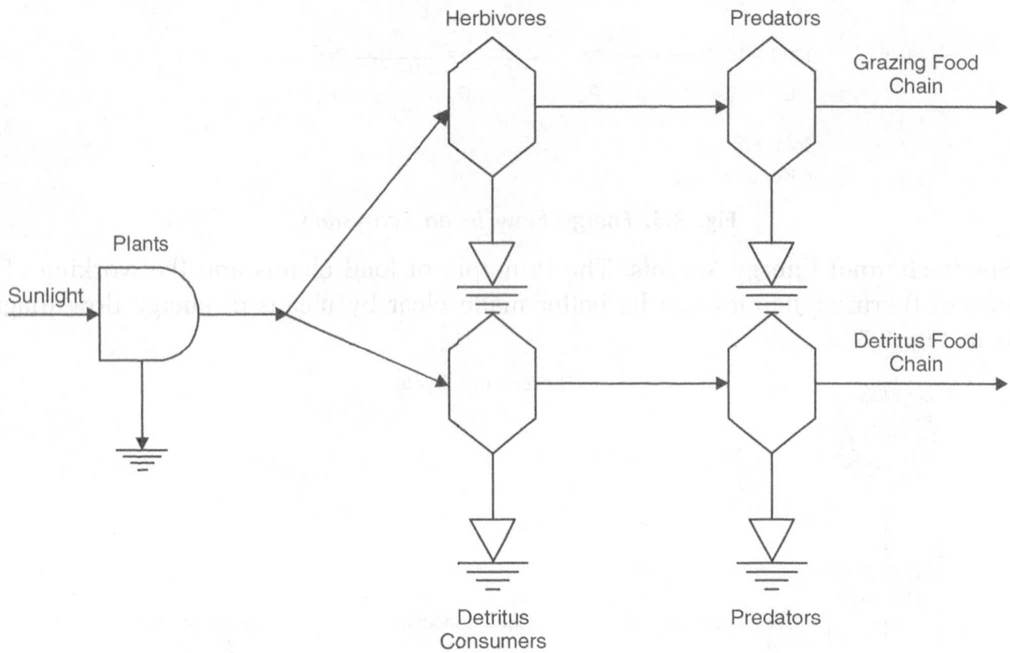

Fig. 3.7. *The Y-shaped Energy Flow Model showing Linkage between the Grazing and Detritus Food Chain.*

1. It confirms to the basic stratified structure of ecosystems.
2. It separates the grazing and detritus food chains in both time and space.
3. The microconsumers and the macroconsumers differ greatly in size metabolism relations.

E.P. Odum (1983) gave generalized model of Y-shaped or 2 channel energy flow model applicable to both terrestrial and aquatic ecosystem.

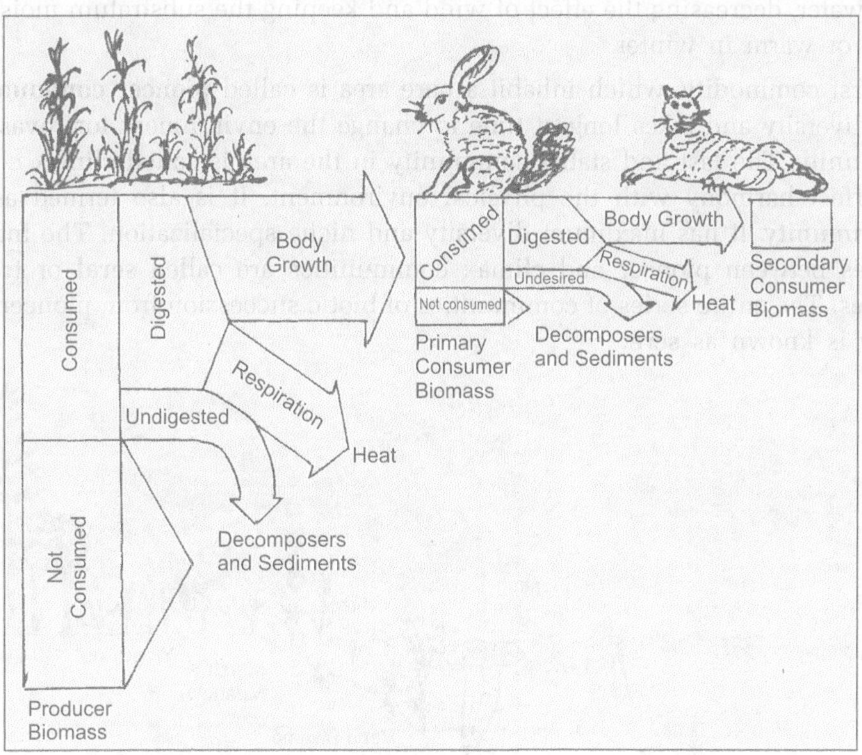

Fig. 3.8. *Flow of Energy through Different Trophic Levels in a Food Chain.*

ECOLOGICAL SUCCESSION

The biotic communities are not stable but undergo regular changes in definite manner with passage of time. Every community undergoes a series of changes until a group of organisms is established, which can live and reproduce most successfully in the area.

"The change of plant and animal community is an orderly sequence in an area, resulting in establishment of stable or climax community is called ecological or biotic succession."

A climax community does not evolve further because it is in perfect harmony with the environment of the area. The interaction among organisms in a community collectively termed biotic factors, influence the structure, composition and function of a community. The physical and chemicals factors of an area determine immediately the kind of community that can be supported there. Bare rocks and shifting sand dunes are unlikely support a forest community due to lack of nutrients and water.

The interacting organisms of the community slowly change the physical and chemical nature of the substratum as well as the factors of the environment. The process of change; is called 'reaction'. It includes weathering of rocky substratum, binding of soil particles, holding of water, decreasing the effect of wind and keeping the substratum moist and cool in summer or warm in winter.

The first commodity, which inhabit a bare area is called **pioneer community.** It has very little diversity and takes longest time to change the environment for invasion of the next community. The last and stable community in the area is called climax community. It is in perfect harmony with the physical environment. It is also termed as **climatic climax community.** It has maximum diversity and niche specialization. The intermediate communities between pioneer and climax communities are called **seral** or **transitional communities.** The entire series of communities of biotic succession from pioneer to climax community is known as **sere.**

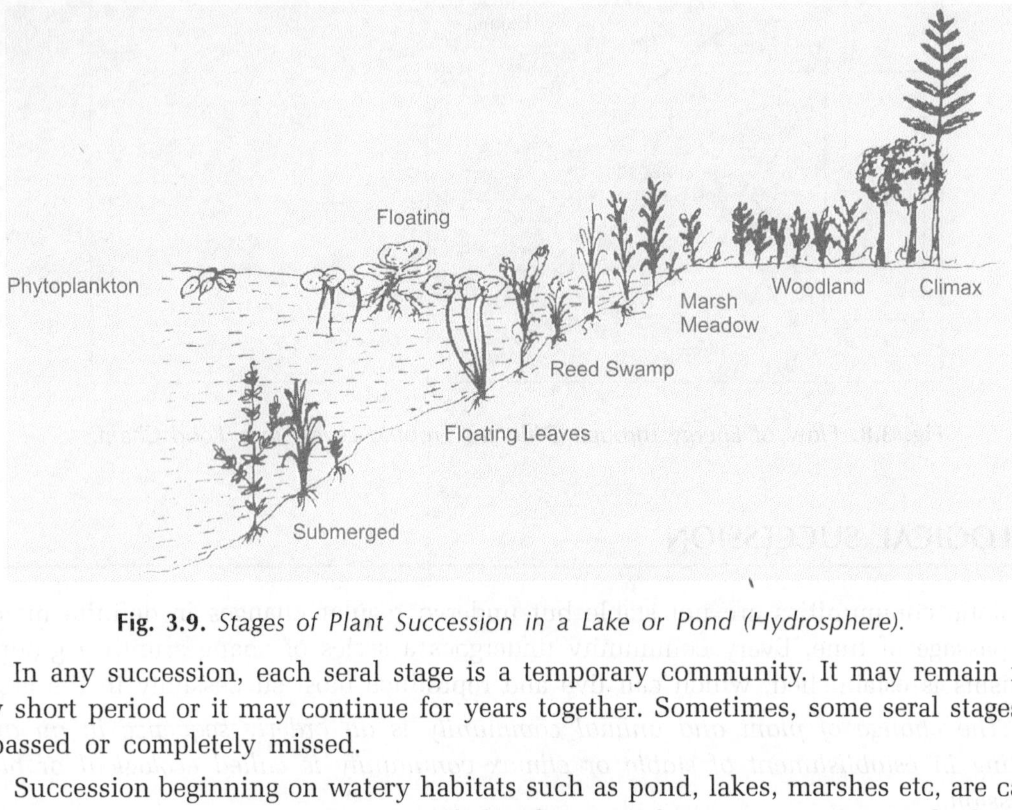

Fig. 3.9. *Stages of Plant Succession in a Lake or Pond (Hydrosphere).*

In any succession, each seral stage is a temporary community. It may remain for a very short period or it may continue for years together. Sometimes, some seral stages are by-passed or completely missed.

Succession beginning on watery habitats such as pond, lakes, marshes etc, are called **hydrarch** and the whole sequence of developmental stages constitutes **hydrosere.** Succession beginning in dry habitats are termed **xerarch** and the different development stages of the succession are called **xerosere.** The xeroseres are of two types:

(*i*) *Lithosere.* Sequence of successional stage on a bare rock and

(*ii*) *Psammosere.* Sequence of successional stages on sand. The climax vegetation in any area is ultimately determined by its climate. The climax community throughout a climatic region is, therefore, the same whether it begins as a xerosere or a hydrosere.

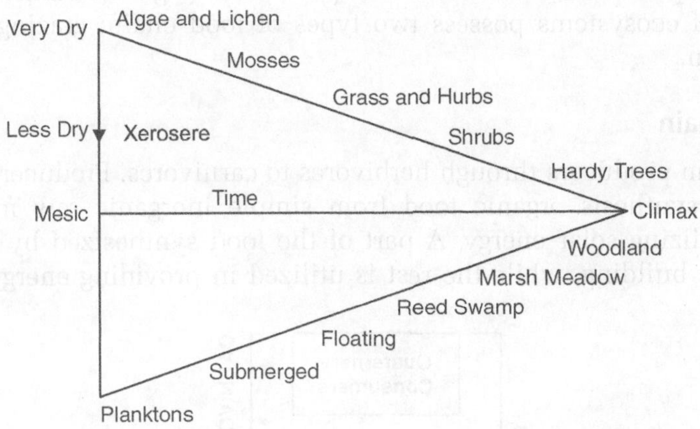

Fig. 3.10. *Convergence in Succession, Xeric or Aquatic to Mesic Vegetation.*

TYPES OF SUCCESSION

Depending upon the nudity of area, biotic succession is of two types:

(*i*) Primary (*ii*) Secondary.

(*i*) *Primary Succession.* The biotic succession that occurs on a substratum devoid of earlier life (*i.e.,* bare rock, lava sediment, new island, newly formed pond or lake etc.,) is called primary succession. The sequence of successional stages of a primary succession is called prisere. Primary succession takes a very long time to reach the climax stage.

(*ii*) *Secondary Succession.* The biotic succession that occurs in an area which have become bare due to destruction of previously existing biotic community by fire, drought, land slide, earthquake etc. is called secondary succession. The sequence of successional stage of secondary succession is called subsere. Secondary succession takes comparatively much less time to reach the climax stage.

Importance of Ecological Succession

The ecological succession is of great practical importance in many activities of man.

(*i*) It provides information, which helps to have a controlled growth on one or more species in a forest by preventing the invasion of other speices in the area.

(*ii*) Pastures can be maintained by not allowing biotic succession to proceed further through grazing or fire.

(*iii*) It also helps in reforestation and forest management programmes.

Food Chain, Food Webs and Ecological Pyramids

In the ecosystem various trophic levels are connected through food chain. The transfer of food energy from the source in plants through a series of organisms with repeated *eating and being eaten* is referred to as a food chain. Or the transfer of energy from one trophic level (*e.g.*, producers) to the next trophic level (*e.g.*, consumers) is called food chain. All types of ecosystems possess two types of food chain; grazing food chain and detritus food chain.

Grazing Food Chain

It extends from producers through herbivores to carnivores. Producers are autotrophic organisms which synthesis organic food from simple inorganic raw materials through photosynthesis utilizing solar energy. A part of the food synthesized by the producers is used in their body building, while the rest is utilized in providing energy for various life activities.

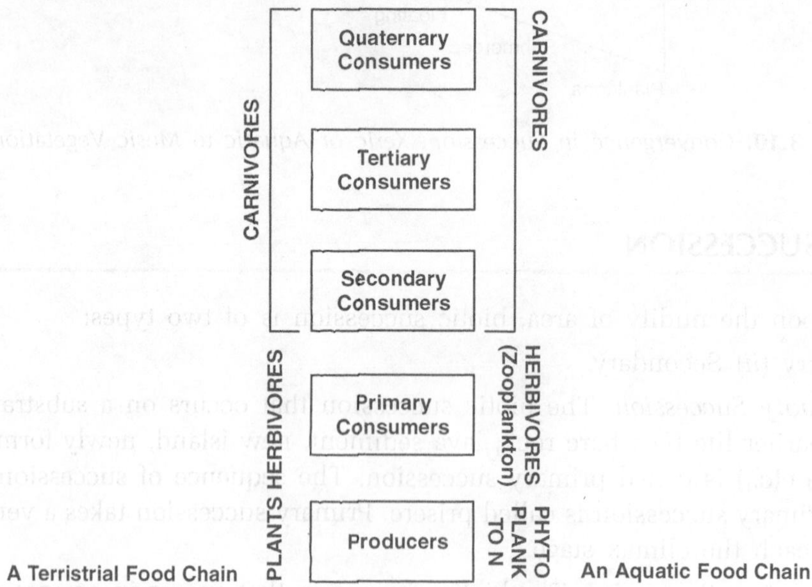

Fig. 3.11. *Food chains and Trophic Levels in Terrestrial and Aquatic Ecosystem.*

Some common food chains are given below:

Terrestrial Food Chains:

(*i*) Vegetation → Grasshopper → Shrew → Hawk.

(*ii*) Vegetation → Rabbit → Fox → Wolf → Tiger.

(*iii*) Vegetation → Frog → Snake → Peacock.

(*iv*) Plant → Butterfly → Frog → Snake → Hawk.

Aquatic Food Chains:

(*i*) Phytoplankton—Zooplankton—Small Crustaceans.

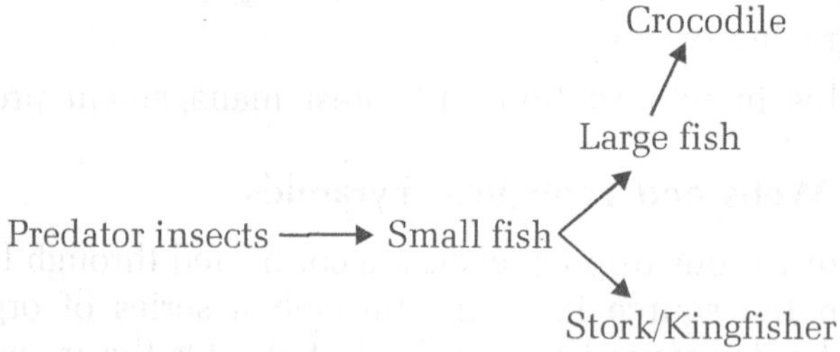

(*ii*) Phytoplankton—Zooplankton—Smallfish—Largerfish—Shark.

(*iii*) Phytoplankton—Zooplankton–Fish–Crane–Hawk.

Detritus Food Chain

It sharts with dead organic matter and passes through detritus feeding organisms in soil to organisms feeding on detritus feeders and their predators *e.g.*, decomposition of accumulated litter in forests. The ecosystems exhibiting detritus food chain are thus less dependent on direct solar energy. These depend chiefly on the influx of organic matter produced in another system. In fact, detritus food chain is simply as sub-component of another ecosystem. A much large fraction of energy flows through the detritus food chain. A good example of a detritus food chain based on mangrove leaves is described below.

Mangrove leaves-Detritus-Micro-organisms-Crabs and shrimps-Small fishes-Large fishes.

FOOD WEB

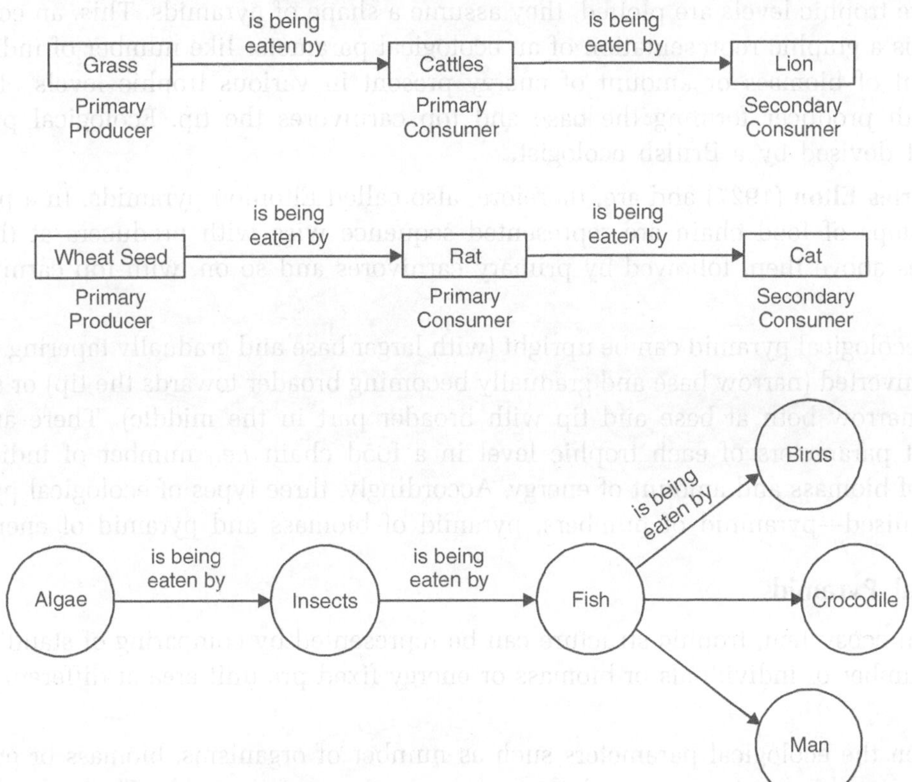

Fig. 3.12. *Simple Food Chain.*

In nature, the food chains are not isolated sequences but are rather interconnected with one another. *'A network of food chains which are interconnected at various trophic levels, so as to form a number of feeding connections amongst different organisms of a biotic community is called food web'.* A food web opens several alternate pathways for the flow of food energy. It also allows an organism to obtain its food from more than one type of organisms of the lower trophic level. Thus, a field mouse may be eaten by a wild cat, a snake or an owl. Similarly, wild cat eats a number of herbivores like birds, mice, squirrels etc. A wolf or jackal can eat both rabbit and deer.

Generally a food web operates according to taste and food preferences of organisms at each trophic level, yet availability of food source and other compulsions are equally important. For example, tigers normally do not eat fish or crabs, but in Sunderbans, they are forced to feed on them in absence of their natural prey.

ECOLOGICAL PYRAMIDS

In an ecosystem, trophic structure can be represented by comparing standing crop (either number of individuals or biomass) or energy fixed pre unit area at different trophic levels.

When the ecological parameters such as number of organisms, biomass or energy at successive trophic levels are plotted, they assume a shape of pyramids. This, an ecological pyramid is a graphic representation of an ecological parameter like number of individuals or amount of biomass or amount of energy present in various trophic levels of a food chain with producer forming the base and top carnivores the tip. Ecological pyramids were first devised by a British ecologist.

Charles Elton (1927) and are, therefore, also called Eltonion pyramids. In a pyramid, various steps of food chain are represented sequence wise with producers at the base, herbivores above then, followed by primary carnivores and so on, with top carnivores at the top.

An ecological pyramid can be upright (with larger base and gradually tapering towards the tip). inverted (narrow base and gradually becoming broader towards the tip) or spindle-shaped (narrow both at base and tip with broader part in the middle). There are three important parameters of each trophic level in a food chain *i.e.,* number of individuals, amount of biomass and amount of energy. Accordingly, three types of ecological pyramids are recognised—pyramnid of numbers, pyramid of biomass and pyramid of energy.

Ecological Pyramids

In an ecosystem, trophic structure can be represented by comparing of standing crop (either number of individuals or biomass or energy fixed pre-unit area at different trophic levels).

When the ecological parameters such as number of organisms, biomass or energy at successive trophic levels are plotted, they assume a shape of pyramids. Thus, an ecological pyramid is a graphic representation of an ecological parameter like number of individuals or amount of biomass or amount of energy present in various trophic levels of a food chain with producer forming the base and top carnivores the tip. Ecological pyramids were first devised by a British ecologist **Charles Elton** (1927) and are, therefore, also called Eltonion pyramids. In a pyramid, various steps of foods chain are represented sequence wise with producers at the base, herbivores above the followed by primary carnivores and so on, with top carnivores at the top.

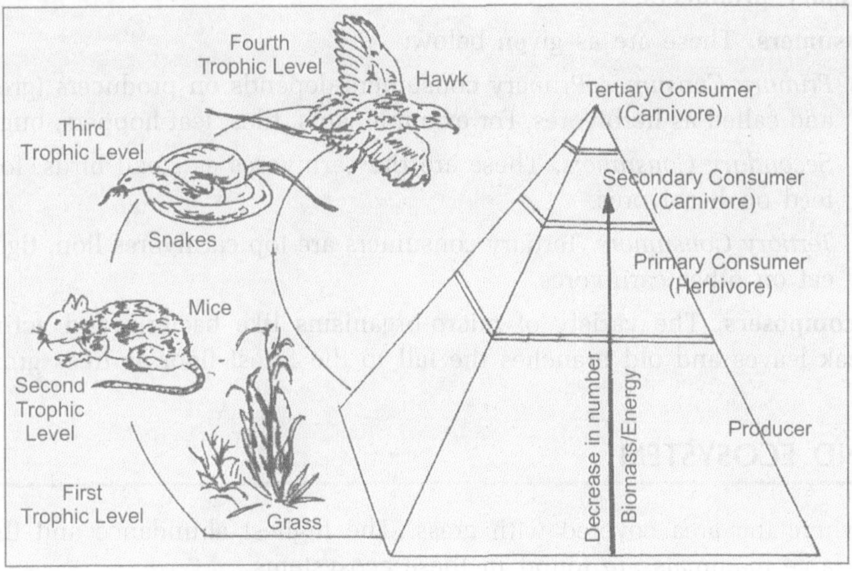

Fig. 3.13. *Formation of an Ecological Pyramid.*

An ecological pyramid can be upright (with larger base and gradually tapering towards the tip), inverted (narrow base and gradually becoming broader towards the tip) or spindle-shaped (narrow both at base and tip with broader part in the middle.) There are three important parameters of each trophic level in a food chain *i.e.*, number of individuals, amount of biomass and amount of energy. Accordingly, three types of ecological pyramids are recognized—pyramid of numbers, pyramid of biomass and pyramid of energy.

FOREST ECOSYSTEM

Forest occupies roughly 40% of the land. In India, the forest occupy roughly one tenth of the total land area. A forest has several types of trees ecosystem either small or large. It depends on type of group species.

Types and Characteristic Features

1. **Temperate Forest Ecosystem.** The temperate forest ecosystem has great importance. Climate where they grow like summer to winter.
2. **The Tropical Rain Forest Ecosystem.** The tropical forest occur in tropical South America, Amazon River basin, East Asia some parts of Africa and northwest.

Structure of Forest Ecosystem

Forest is very complex ecosystem, because there are so organisms. They provide food for other organisms, while provide shelter or control populations through predation.

(*a*) **Producers.** In a forest ecosystem trees or plants are the source of energy. Where they capture the sunlight and forms is food. In this way plants are called the **primary producers.**

(*b*) **Consumers.** These are as given below:

(*i*) *Primary Consumer.* Primary consumers depends on producers (green plants) and called as herbivores. For example, ants, tiles, leaf hoppers, bugs deer etc.

(*ii*) *Secondary Consumers.* These are the carnivores as small birds, fox etc. and feed on herbivores.

(*iii*) *Tertiary Consumers.* Tertiary consumers are top carnivores lion, tiger etc. that eat on other carnivores.

(*c*) **Decomposers.** The variety of micro-organisms like bacteria and actinomycetes break leaves and old branches the fall to the forest floor as trees grow.

GRASSLAND ECOSYSTEM

Grassland where the area covered with grass. The highest abundance and the greatest diversity of large mammals are found in these ecosystems.

Structure and Function

The various components of a grassland ecosystem are as follows:

1. **Abiotic Component.** The abiotic components are the nutrients present in soil and the aerial environment. Some elements like H, O, N, P etc., are supplied by CO_2, phosphates and sulphates etc., are present in air and soil of ground.
2. **Biotic Component.** These may be categorised as:

(*a*) **Producers.** In a grassland ecosystem, the producers are mainly grasses, birds and shrubs also participate to primary production.

(*b*) **Consumers.** These are as given below:

(*i*) *Primary Consumer.* The herbivores directly depends on success for example, cows, deers, sheep, rabbit etc. Besides them, there are present some insects, termites and millipede etc., that feed on leaves of grasses.

(*ii*) *Secondary Consumers.* The secondary consumer are the carnivores that feed on primary consumers, such as fox, snakes, frogs, lizards etc.

(*iii*) *Tertiary Consumers.* The tertiary consumers that feed on secondary consumers like hawks etc.

(*c*) **Decomposers.** The microbes breakdown the decay of dead panic matter of different forms of higher life are fungi as species Mucor, Penicillium, Rhizopus etc. They bring about the miner-back to the soil, thus making them available to the producers.

DESERT ECOSYSTEM

Deserts are recognised by scantly flora and fauna. Deserts occurring where annual rainfall of less than 25 centimeters. Temperature and water factors affect the species composition.

Structure and Function

The different biotic components are given below:

(*a*) **Producers.** In a desert ecosystem producers are shrubs, especially bushes, some grasses and a few trees. Sometimes a few succulent like cacti are also present. Other lower plants like herbs and xerophytic mosses may also be present.

(*b*) **Consumers.** The most common animals like reptiles and insects, live under xeric conditions. There are also find some natural rodents and birds. Camels feed on tender shoots of the plants and called as 'Ship or desert'.

(*c*) **Decomposers.** Due to poor vegetation the amount of dead organic matter is correspondingly less. In this way these are very few numbers. Decomposers are some fungi and bacteria, most of which are thermophilic.

AQUATIC ECOSYSTEM

Aquatic ecosystem includes fresh water and marine water ecosystem.

Fresh water ecosystem includes lentic and lotic system. Marine water ecosystem includes open ocean and coastal ecosystems.

(*a*) Pond Ecosystem

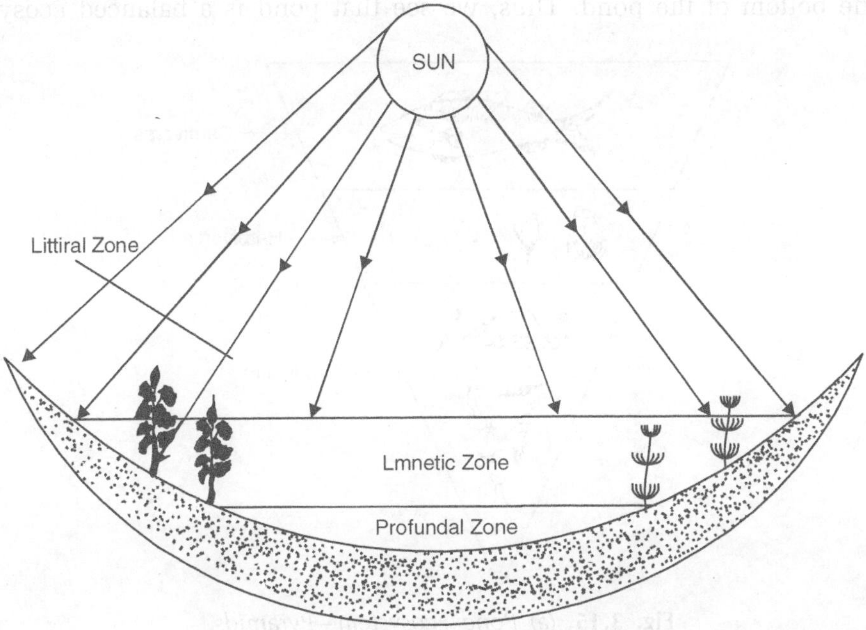

Fig. 3.14. *Different Zones of a Fresh Water Pond.*

The ecosystem is the basic functional unit in ecology. Since it includes both the biotic communities and abiotic environment, each influencing the properties of the other and both are necessary for maintenance of life as we see on earth. A pond or lake is a classical example of the ecosystem.

Let us consider the pond as an ecosystem. It is not only a place where plants and animals live but plants and animals make the pond what it is. It can be divided into four basic unit:

(*i*) *Abiotic substance.* These are water (H_2O), Carbon dioxide (CO_2), Oxygen (O_2), Calcium (Ca), Nitrogen (N), Phosphorus (P), amino and humic acid etc. A part of these vital nutrients is in solution and is available to organisms while much larger portion is held in reserve in particulate matter.

(*ii*) *Producers in a pond.* Producers in a pond are of two types: (*a*) rooted or large floating plants generally growing in the shallow water, (*b*) minute floating algae called phytoplankton-distributed as deep as light penetrates. When in abundance phytoplankton give greenish colour to the water. They are more important in the production of basic food for the ecosystem.

(*iii*) *Consumers in a pond.* Animals such as insects, larvae, crustaceans and fish feed directly on living plants or their remains and are called primary consumers. Whereas the carnivores *e.g.*, carnivorous fish that feed on the primary consumers are called a secondary consumers.

(*iv*) *Decomposers.* We find aquatic bacteria and fungi distributed throughout the pond. When temperature conditions are favourable, they decompose the dead organisms and the materials are released for reuse. The entire pond can be divided into two strata—An upper 'production zone' and a lower, 'decompositions-nutrient regeneration zone'. Organic detritus is consumed and decomposed on the bottom of the pond. Thus, we see that pond is a balanced ecosystem.

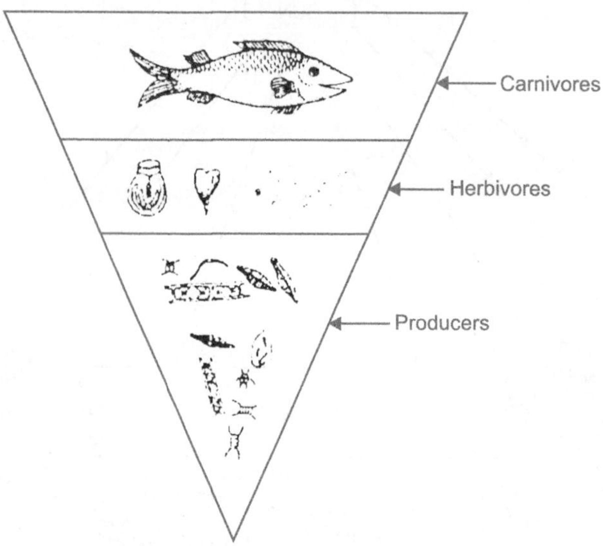

Fig. 3.15. *(a) Pond Ecosystem—Pyramids.*

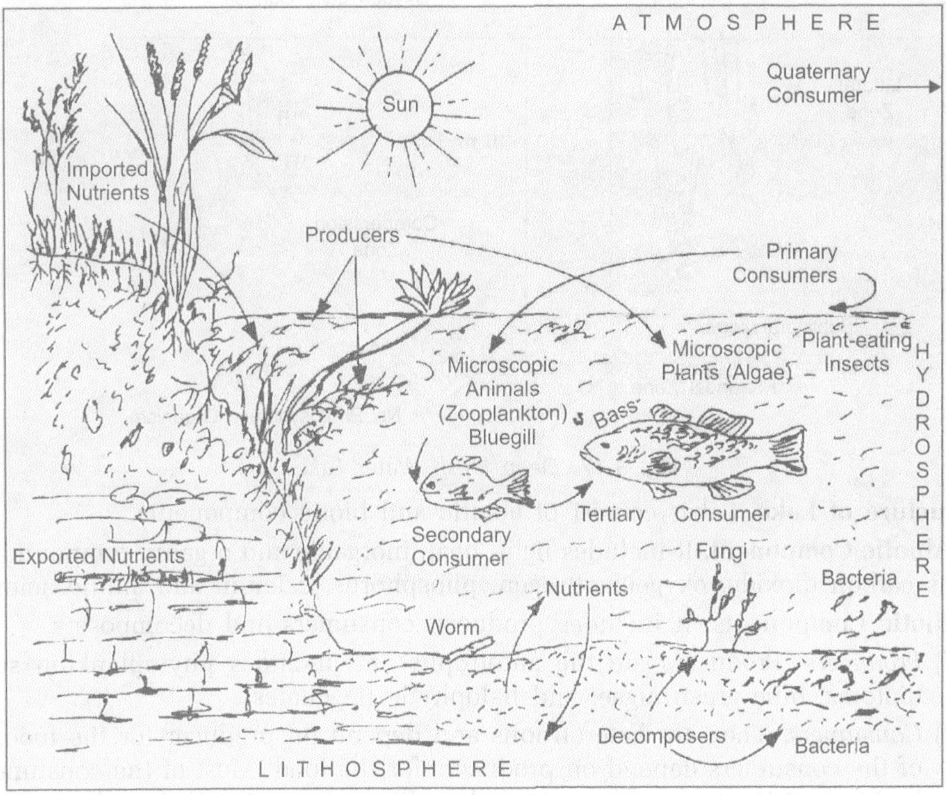

Fig. 3.15. *(b) Pond Ecosystem.*

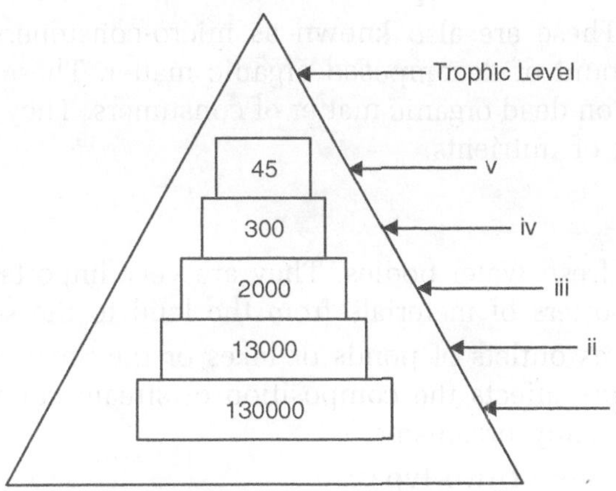

Fig. 3.16. *Biomass Production at Various Trophic Levels in the World.*

(*b*) Lake Ecosystem

The abiotic component of lakes depend on location, latitude, altitude and the surrounding ecosystem. The depth ranges from few feet to 5000 feet. The gradation of oxygen, light and temperature affect the life in the lake. Lake are two types.

(*i*) Oligotrophic lakes.

(*ii*) Entrotrophic lakes.

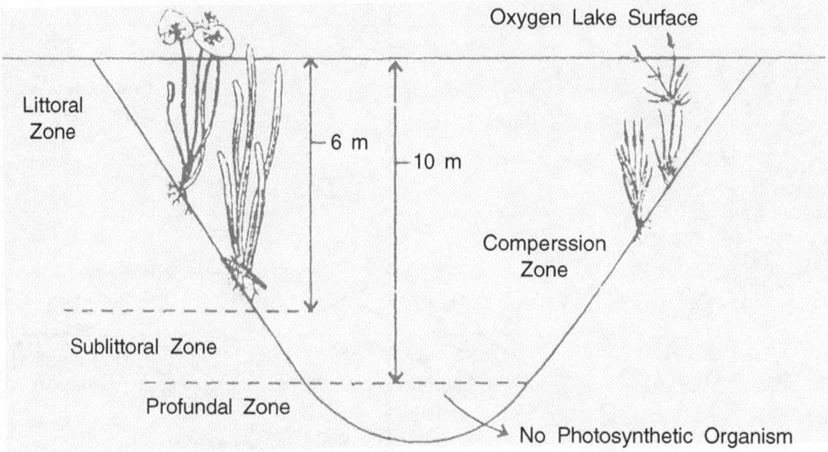

Fig. 3.17. *Deep Fresh Water Lake.*

Structure of Lake. Lakes consist of abiotic and biotic components.

1. **Abiotic Component.** It includes light, heat, inorganic and organic compounds such as carbon dioxide, oxygen, nitrogen, phosphorus, calcium and amino acids.
2. **Biotic Components.** It includes producer, consumers and decomposers.
   (*i*) *Producers.* Producers are the autotrophs and includes phytoplankton such as diatoms, blue green algae, and halophytic flagellates.
   (*ii*) *Consumers.* They are heterotrophs and depend on producer for the food. Most of the consumers depend on producers for the food. Most of the consumers are herbivores. Here a consumer includes oligocheates, Autodrilus plurisesta, Bran chira sowerbyi, hirudinea species, molluscs and crustanceans etc.
   (*iii*) *Decomposers.* These are also known as micro-consumers because they absorb only small amount of decomposed organic matter. These includes bacteria and fungi that feed on dead organic matter of consumers. They also play an important role in cycling of nutrients.

(c) Stream

Stream are flowing fresh water bodies. They are very important in the biosphere as they act as major transporters of materials from the land to the sea/ocean.

Streams may begin as outlets of ponds or lakes or they may arise from springs and seepage areas. Temperature affects the composition of stream community, shallow water stream are not good for many organisms.

Types of Streams: These are of two types:

1. **Fast Streams.** It has high velocity of water and makes excellent trout nurseries.
2. **Slow Streams.** It has low velocity of water, silt and decaying organic matter accumulates on the bottom.

Structure of Streams

1. **Abiotic Component.** These includes inorganic organic substance and climatic factor.
2. **Biotic Component.** Biotic Component of streams are as follows.
   (*i*) *Producers.* It includes phytophanktons like diatoms, blues green algae and green algae like cladopora, ulotherix and water moss.

(*ii*) *Producers.* its includes cephatophexy Helodes, Phalacrocera, Gammarus etc. Fishes are also present.

(*d*) River

River includes moving water. They are variable and ranging in size from Ganga, Yamuna etc. A river varies considerably over its length, as it changes from a Mountain brook to a large river.

1. **Abiotic Components.** These include inorganic, organic substance and climatic factors.
2. **Biotic Components.** Biotic components of river are as follows.

(*i*) *Producers.* Producers includes phytoplanktons, diatoms, blue green algae etc.

(*ii*) *Consumers.* Consumers includes cephatopheyx, Helodes, Phalacrocera, Grammarus etc.

(*ii*) *Decomposers.* It includes bacteria and fungi which feed on dead organic matter.

(e) Oceans

The oceans of the world cover 70% of the earth's surface and they occupy a volume 15 time more than that of the land. The oceanic environment is very different from that of inland water, the major difference being the chemical composition of water.

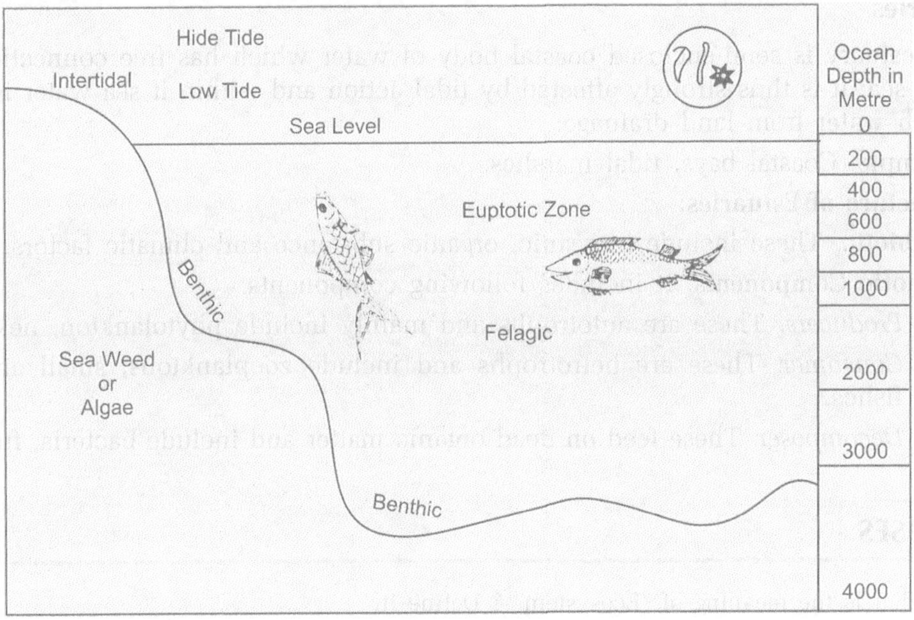

Fig. 3.18. *An Ocean Ecosystem.*

Structure of Ocean. It consist of abiotic and biotic components.

1. **Abiotic Components.** It includes inorganic and organic substances. Climatic factors are also the part of abiotic components.
2. **Biotic Components.** It includes producers, consumers and decomposers.

(*a*) **Producer.** It includes autotrophs. Producer are phytoplankton such as dinoflagellates, diatoms and algae. In case of algae, phaeophyceae and rhodophyceae are the dominating ones. All contributes to primary production.

(*b*) **Consumers.** These are heterotrophs macro consumers. They depend on primary producers for their nutrition.

(*i*) *Primary Consumer.* These include herbivores. They feed on producers–mainly, fish etc.

(*ii*) *Secondary Consumer.* These are the carnivores. They feed on herbivores. Example: Secondary consumers are fish such as shad etc.

(*iii*) *Tertiary Consumer.* These are also the carnivores. They feed on carnivores of secondary consumers. These mainly involve fishes such as cod, hadlock etc.

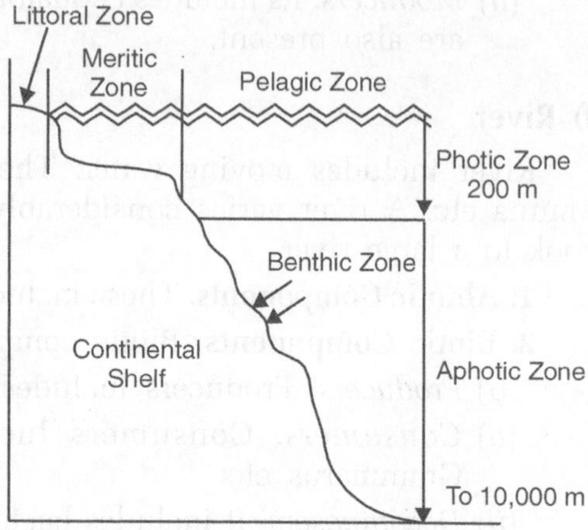

Fig. 3.19. *Horizontal and Vertical Zonation in the Sea.*

(*iv*) *Decomposer.* These are bacteria and fungi. They decompose dead organic matter of producer. They aslo help in cycling of nutrients by decomposition.

(*f*) Estuaries

An estuary is semi-enclosed coastal body of water which has free connection with the open sea; it is thus strongly affected by tidal action and within it sea water is mined with fresh water from land drainage:

Example. Coastal bays, tidal marshes.

Structure of Estuaries:

1. **Abiotic.** These include inorganic, organic substance and climatic factors.
2. **Biotic Components.** It includes following components.

(*i*) *Producers.* These are autotrophs and mainly include phytolankton, nektons.

(*ii*) *Consumer.* These are hetrotrophs and include zooplanktons, small and large fishes.

(*iii*) *Decomposer.* These feed on dead organic matter and include bacteria, fungi etc.

EXERCISES

1. What is the meaning of "Ecosystem" ? Define it.
2. Write different types of ecosystem.
3. Write detailed note on "Ecological Succession".
4. Write short note on Food Chain.
5. Discuss a short note on Food Web.
6. Write a detailed note on "Ecological Pyramids".
7. Write short notes on:

(*a*) Grassland Ecosystem (*b*) Forest Ecosystem

(*c*) Desert Ecosystem (*d*) Aquatic Ecosystem

(*e*) Ocean Ecosystem.

❑❑❑

Unit IV BIODIVERSITY AND ITS CONSERVATION

BIODIVERSITY

"Biological Diversity or Biodiversity" is that part of 'nature' which includes the difference in genes among the individuals of species, the variety and richness of all the species in a region, as well as the various types of ecosystem within a defined area."

All these levels of variability are a part of the living components of our planet called the 'biosphere'. The unit of biodiversity referred to, could be global or sub-global, or could form a characteristic of geographical region, or a political entity such as a country, as state.

An ecosystem is a geographical unit with its own biotic features such as climate, soil, water and a variety of interlinked species of plants and animals. Examples are the forests, grasslands, deserts, river, lakes and seas around us. Each ecosystem has its own level of biodiversity that is measured by the number of species present in it.

The biodiversity of a region is a valuable' natural resource', which forms a part of every ecosystem. Man is a part of these ecosystems and is dependent on their integrity. Diversity in wild species form the 'gene pool' from which our crops and domestic animals have been developed over thousand of years. Today the variety of nature's bounty is being further harnessed through breeding programmes for crops and domestic animals or through biotechnology, by manipulating genes for developing better types of medicines and industrial products.

Food, clothing, housing, energy, medicines, are all resources that are directly or indirectly linked to the biological variety present in the biosphere. The biosphere forms a relatively thin crust in which life is found on our planet.

CLASSIFICATION OF BIODIVERSITY

Biodiversity deals with the variety of living organisms found in the biosphere. This variability can be observed at three levels :

1. **Genetic Diversity.** This is the variability among individuals of a species *e.g.*, each human being differs widely from all others. This diversity is due to the large

number of combinations possible in our genes that give us specific characteristics. If the number of breeding individuals is reduced, inbreeding occurs. Eventually it can lead to the extinction of the species.

Table 4.1

Taxonomic Group	*Number of Species*
Higher plants	2,70,000
Algae	40,000
Fungi	72,000
Bacteria	4,000
Viruses	1550
Mammals	4650
Birds	9700
Reptiles	7150
Fish	26959
Amphibians	4780
Insects	1,025,000
Crustaceans	43000
Molluscs	70,000
Nematodes and Worms	25,000
Protozoa	40,000
Others	1,10,000

Approximate numbers of species from major taxonomic groups which have been described from all over the world.

2. **Species Diversity.** The number of species of plants and animals that are present in a region constitutes its species diversity. This diversity of species is seen both in various natural systems, and in agricultural systems. Natural tropical forests have much greater species diversity than most other regions. Natural forest has higher species diversity than a timer plantation, which is usually a mono-culture. In the natural system there are a large number of non-wood products that people depend on such as fruit, fuelwood, fodder, fiber, gum, resin and medicines. Timer plantations do not provide such goods for local consumption.
3. **Ecosystem Diversity.** Each area has a variety of different ecosystems. These have their own complement of distinctive interlinked species. Each of them varies functionally and structurally from other system. The ecosystem in an area is referred to as 'natural' when it is relatively undisturbed by human activities or 'modified' when it is changed to other types of uses. Natural ecosystems are forests, grasslands, deserts and aquatic systems, either fresh water or salt water. These are wilderness areas with their own wild plants and animals, which are of great economic value. If they are overused or misused, their productivity eventually decreases. The ecosystem is then said to be degraded.

Biogeographical Classification of India

India is rich in with a great percentage of endemic flora and fauna. The main reason for its variety of climatic and altitudinal conditions with different ecological habitats. The other some countries of mega diversity are Australia, China, Peru, Mexico etc. India has some different biogeographical regions like the Indo-Gangetic Plain, peninsular Indian subregion, The trans-Himalayan, Desert, Andaman and Nicobar islands, Semiarid zone etc. But according to the distribution and environmental interaction of flora and fauna it has been classified into 10 biogeographical region. Each region identified by some important properties like biodiversity, soil profile, topographic factors, climatic conditions etc. The flora and fauna quite differ from one region to another region. Further division into 12 regions is identified by characteristic plants and animal species. These 12 regions known as **biogeographic regions.** But 10 biogeographic regions has characteristic and rare species of flora and fauna. Some are not available anywhere in the world.

Andman and Nicobar Islands

Fauna. A great number of reptiles and marine animals occur in this region. About 250 bird species are found there. In which megapede, white breasted switlet, Nicobar pigeon and different fruit pigeons are commonly found here and a man pig, palm civet, barking deer, sambar etc., are found at a large amount. Pythons, viper, coral and sea snake, cobras are some variety of snakes found here.

Flora. There are some species which dominantly occur like, *Lager stroemia, Terminalias* etc. Coastal areas are rich due to mangroove vegetation.

VALUE OF BIODIVERSITY

Daily requirements of human beings are fulfilled by biodiversity resources. Biodiversity contains a wide range of animal and plant species and used for several purposes like manufacturing of food, cosmetic products and pharmaceutical. Different types of animals and plants have a economical value. In this way we are directly or indirectly depend on biodiversity. The value of biodiversity is divided into two parts :

1. Direct values which are further divided into two:

   (*a*) Consumptive use value, (*b*) productive use value.

2. Indirect values : Non consumptive uses.

1. Direct Values

(*a*) Consumptive Use Value. Where the products are consumed directly. For example- fuel wood, food, medicines, leather, rubber, drugs, silk etc.

(*i*) *Food.* There are a large number of food products which are consumed by rural populations. Wild plants like leaves, roots, tubes, mushrooms, variety of fruits, nuts, coffee, etc., highly consumed. Wheat, maize and rice gives about 60% calories and 55% protein comes from plants into the humans.

(*ii*) *Medicines.* Medicine play a crucial role in the medication. Several types of medicines extracted from higher plants. There are a wide range of wild species who gives us medicinal benefits. Medicinal plants are important for Ayurveda and allopathic treatment. Except of plant a large number of hosts also play a vital role in medical treatment. Micro-organisms provide life saving drugs like penicillin, antibiotics etc.

(*b*) **Productive Use Value.** Productive use value are related to commercially usable values, in which the products is come to the market and then sold. In this way they have reproductive use value. Several industries are associated to productive use value of biodiversity.

2. Indirect Values

1. **Social Values.** Social values are related to social aspects of human beings like social life, religion, customs, etc. There are several groups of plants, which are used in worship.
2. **Ethical Values.** Ethical values is known as existence value. Education, rights and duties are ethical views of biodiversity is to be shown relationship with the other living species.
3. **Aesthetic Values.** Biodiversity has a big aesthetic value. Where wild animals are the main attraction of beautiful natural landscapes and developed into a new industry as tourist industry.
4. **Option Values.** These are unknown values and need to be explored. These industries are directly related to micro-organisms, genes, molecules etc. The pharmaceutical field is more interested in wild species.

Biodiversity at Global, National and Local Levels

Number of animal species in India and the world are not evenly distributed. About 90% fish, 85% higher plants, 98% birds and 95% reptiles are exist on the earth. Most of the species make up by insects. Our trophics are rich in biodiversity, because favourable conditions for evolution and minimum extinction, interaction between organisms, topography, heredity, climate and time.

National Level. India has a large land area about 3,287,263 sq.km. The tropical waters of the Indian ocean with the Bay of Bengal lying to the South-East and the Arabian sea to the South-west. About 40% land area used for cultivation. Insects cover half of the biodiversity. These occur in ocean water and on land with parasitic or mutualistic state. In the world 16,04,000 species of Fungi, Animalia, Protista occur on the land. The country has a great amount of biodiversity with a sizeable percentage of flora and fauna. India has three biomes like dry desert, tropical dry forests and tropical humid forests and two realms as the Indo-Malayan and Palaeoarctic. Out of 25 Hot spots known but India has only two, one is North-East India and another one is Western Ghats. Some crops like jute, rice, banana, mango, citrus, sugarcane are first grew in our country and spread throughout the world. In this way origin occur in terms of agro-biodiversity.

Local Level. In this level biodiversity is characterized by grouping species. Alpha diversity where number of species found in a small homogeneous and Beta diversity where changes occur between species composition across different habitats. Gamma diversity means a rate of change across large geographic areas. Point diversity as the number of species occur at a single point in a given space.

Genetic Diversity in a Rabbit Population

⇩

Species Diversity in a Prairie Ecosystem

⇩

Community and Ecosystem Diversity Across the Landscape of an Entire Region

Different Levels of Biodiversity

India as a Mega-Diversity Nation. Our country is famous for the variety of wildlife occur here. But some years past wildlife has depleted rapidly. About 38 species of birds, 66 of mammals and 18 species of reptiles and amphibians are rare and endangered. These have been recorded in Wildlife Act, 1972. Cheetah a wild animal has become extinct, 8.6% of wild animal and plant species cooperate the global biodiversity of India. But there are some specific reasons reduces its stability like modification, degradation and loss of their habitats. So due to this our flora and fauna are comes under threat of extinction. Some other reasons also play a key role of its extinction such as over exploitation, introduction of exotic species, urban housing, complexes, pollution and global warming. Along 7500 km, long coastline has a rich biodiversity. But there are a great number of species awaiting discovery. India has tenth number of the world countries and fourth in Asian countries. There are 17,000 species of flowering plants in which 35% are endemic. Indian is one of the 12 mega-diversity centers of the world.

Today the extinction rate is high over the conservation. So there are a great need for their protection and appropriate steps are taken for its conservation.

Hot Spots of Biodiversity

Human activities are responsible for the degradation of hot spots area. Where several endemic and threatened species live. Out of 25 hot spots, two are found in the Indo-Burma region and the western Ghats. These are rich in floral wealth and also in amphibians, reptiles, butterflies and some mammals.

Hot spots support rich biodiversity because of endemic flora and fauna, geologic formations and other scientific interest. There are 12 regions which have been classified as Mega-biodiversity centers these regions which have been classified as Mega-biodiversity centers these regions are India, China, Peru, Brazil, Mexico, Malaysia, Australia, Indonesia, Columbia, Zaire, Ecuador and Madagascar.

Myers in 1988 has identified 10 hot sports of tropical forests. These hot spots cover 0.2% land area. In 1991 Myers added 8 areas. Two regions of India, Western Ghats and Eastern Himalayas are known as hot spots. There are some global hot spots are given below : (1) Western Ecuador, (2) Madagascar, (3) Tanzania, (4) New Caledonia, (5) Philippines, (6) Eastern Himalayas (India), (7) Borneo (North), (8) Western Amazonia, (9) Peninsular Malaysia. (10) Californian province.

Western Ghat are the major genetic estate with an enormous biodiversity. There are two main centers of diversity, the silent valley and the Agastyamali Hills.

Eastern Himalayas has forest cover from 340,00 km^2 to 110,000 km^2. But today only 6.8% vegetation exists according to Myers. There are some commonality occur between Western Ghat and Eastern Himalayas. Some faunal species like Lizard, Hawk, Flue bird, Hornbill are found in hot regions.

Threats to Biodiversity

Threat to biodiversity is an important issue in present scenario, on each there is no place where there is no ecological catostrophe. Due to which one-third of the species are likely to extinct in the coming year. The tropical forest are shrinking at the rate of 0.8 percent each year.

Causes of Extinction of Biodiversity

There are several reasons for the extinction of Biodiversity.

1. **Hunting.** Hunting applied for several purposes like hunting for enjoyment, hunting for money etc. Hunting is a quick and easy way for earning money. A tiger skin has great value about thirty five thousand rupees. Similarly, the tusks of elephants also so costly. Several snake species reduced due to export of their skin. The demand of many animals like wild bear, nilgai and deer consumed by many peoples. Hunting is a great fun for several rich people of India. The number of animals shot for pleasure with modern aids like binoculars, jeeps etc. But sometimes our hunter used some bad hunting methods like use of poison results the death of many animals. On the other way forest fire for driving animals in one direction for being shot.
2. **Habitat Loss.** Wild animals require shelter for their survival in the form of forest vegetation, grassland, river margins, etc. But absence of these covers animals will not survive even if water and food available in excess.

   Habitat loss occur due to pollution when river and stream run through the forest, many factories gives affluent which pollute the river and streams water which is unsafe for animals.

   In other way destruction of wild plants which is main source of food for several animals, cause its depletion.

   Building roads and railways also create a problem. When rail pass through the dense forest, the timid animals afraid from noise. Forests areas are also converted into the crop lands. These are the main reasons for the freedom of movement of the wild animals reduction.
3. **Man Activity.** Human being diminish our biodiversity at an alarming rate. Activity of man main cause for their depletion. When man do not improve in their nature, the rules and regulations may not helpful.
4. **Over Exploitation.** Over exploitation is one another important cause for wildlife destruction. Over use of wild plants leads to commercial exploitation. Some plants which has high medicinal value and horticultural plants comes under the over exploitation.

5. **Other Factors.** Here some other factors are mentioned which helpful for the extinction of wildlife like—Loss of gene flow, out breaks of diseases, reproductive rate, distribution range, organism position in the off chain specialization degree etc.
6. **Poaching of Wildlife.** Poaching is another insidious threat that has emerged in recent decades as one of the primary reasons for the decline in number of species. Despite legal protection in many countries, products from endangered species are widely traded within and between nations.

   Wildlife is old for live specimens, folk, medicines, fears, hides, skin and other products such as ivory, horns amounting to millions of dollar each year.

MAN-WILDLIFE CONFLICT

Man and wildlife conflict is manifest between local communities and the wildlife protecting bodies. The protection of wildlife area is an indirect conflict with the economic interest of local communities. The rural communities have not only lost their wildlife resources, but they also bear the full costs of crop damages caused by migrating wildlife.

An example of man-wild conflict is human elephant conflict. The elephant commands the affection of the people. Elephant are intelligent, lovable for the majority of the human inhabitants, they are the creatures to be admired and cherished. However, once you put some of these elephant next to your house where they could not only rob you, your food, your livelihood and possibly your life, attitude swiftly changes to conflict.

Endangered Species of India

Endangered species are those who have a high likelihood of "going extinct in the near future." Following animals have been declared endangered and have been declared protected species by the Government of India :

(1) Elephant, (2) Muggar, (3) Leopard cat, (4) Lion, (5) Chital, (6) Golden cat, (7) Musk deer, (8) Indian bison, (9) Asiatic wild ass, (10) Indian rhinoceros, (11) Crocodile, (12) Golden langur, (13) Pigmy hog, (14) Markhor, (15) Red jungle fowl, (16) Kashmir stag, (17) Snow leopard, (18) White tiger, (19) Slender lories, (20) Indian gharial.

There are some reasons for their extinction :

1. Export of some species.
2. Poaching for fur, meat, skin, rhinohorns etc.
3. Overgrazing by domestic animals.

About 1000 creatures are threatened with extinction, according to the Red Data Book of IUCN (International Union for Conservation of Nature and Natural Resources).

About 450 plant species as endangered, threatened or rare.

For the endemic fauna the Western Ghats is famous. Nilgiri Langur, Jungle cat, golden cat, lion tailed macaque etc. and some important endangered species. Himalayas musk deer are known as endangered due to over-expectation. Today elephant is considered as an endangered species and is limited to geographical zones. The musk deer is a threatened species. Musk deer became as an endangered species due to indiscriminate killing by poachers in Kashmir, Sikkim, Ladakh and Garhwal. Cheetah found in central India, Rajasthan, Jaipur and Deccan, but unfortunately it is an extinct species in our country. Leopard was once widely distributed in the country, but its number has come down drastically due to over hunting. Snow leopard found on Himalayas and is a beautiful animal with white grey colour and large black rings but it was overhunted for its fur and is known as endangered species. The lion tailed macaque is an indangerous species and found in the Cardamom hills and Wildlife Sanctuary. The number of lions declined due to killing by man and deforestation, because it is being hunted for its fur and meat. The Great Indian Bustard is an endangered bird species and is now being protected in the Thar Desert Sanctuary.

Endemic Species of India

According to world's recorded endemic species, about 3,501 one found in the Himalayas and 1601 in the Western Ghats alone. The endemic species found in large number in our country. About 33% endemic flora found in the North West Himalayas, Western Ghats and the Andman and Nicobar islands. About 60% amphibian species are endemic and found in Western Ghats.

CONSERVATION OF BIODIVERSITY

Biodiversity can be conserved in two ways *e.g.*, *ex-situ* conservation and *in situ* conservation. A decade ago, biodiversity conservation was limited to saving genes, species and habitats. But now a new conservation framework has emerged based on saving biodiversity, studying it and using it sustainably. The steps needed go far beyond maintaining protected areas and seed banks to include reforming, agricultural, forestry and technological policies and revamping international trade agreements and watershed management. Within this frame works, countries at all levels of development as well as scientific, new government organizations and the private sector can see the relevance of biodiversity to their needs. Different activities can therefore, be linked to biodiversity conservation and use.

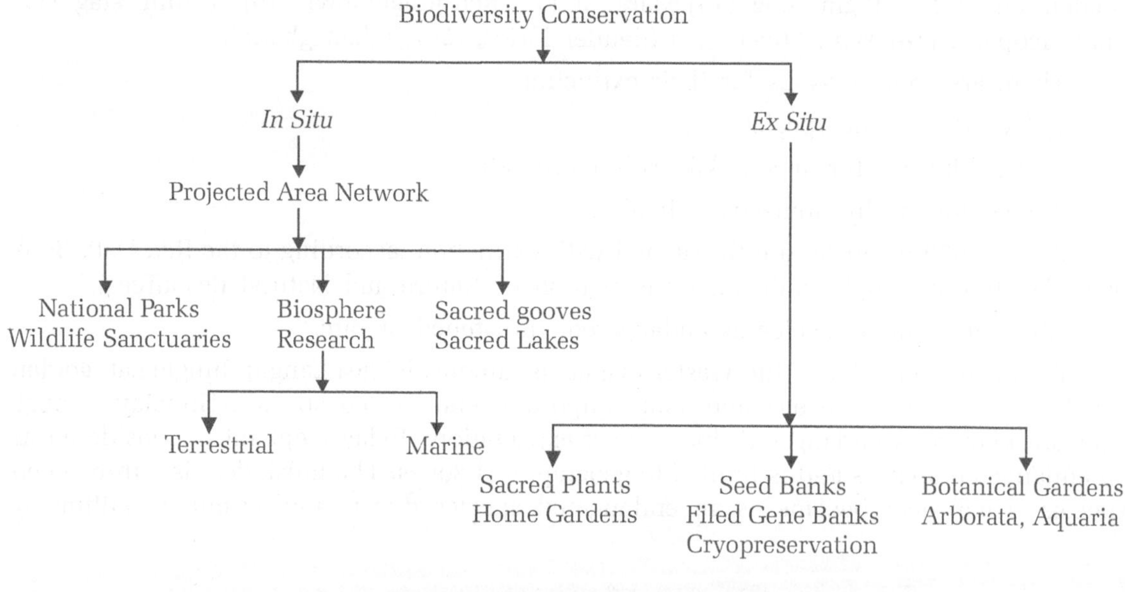

1. *In-situ* Conservation

In-situ conservation means preservation of species in its natural ecosystem. There are different areas, which are managed for *in-situ* conservation. These are given below :

(*a*) **National Parks and Sanctuaries.** National Parks are an areas for protection of flora, fauna and landscape and their conservation. But in sanctuary, protection is given only to animals. There are 368 sanctuaries and 66 national parks have been set up in India. Both these areas are being declared by Wildlife Protection Act 1972.

(*b*) **Special Projects**

(*i*) *Project Tiger.* In our country there were 40,000 tigers in the year 1910. Project tiger was started as a Central Sector Scheme in 1973. At the starting time there were 9 tiger reserve located in 9 state but two more reserves were added in 1982 and making the total tally to 11 reserves with, an area of 156,800 sq. km. in 10 states.

(*ii*) *Gir Lion Projects.* Five-year plan prepared by the Gujarat Government for the protection of lion. Excessive cattle grazing and clearing of forest lead to decline in the lion population. There are 240 lions in the Gir forest according to 1985.

(*c*) **Biosphere Reserves.** A biosphere reserve contain domestic animals, wild animals, cultivated plants, etc. and is regulated by the government. Biosphere reserve is formed for multiple land use. It has different types of zone like—Restoration zone, Manipulation zone, Buffer zone and Core zone. A Biosphere Reserve Program was launched by UNESCO in 1971 under its "Man and the Biosphere Program" (MAB). The objects of this programme given below :

1. To provide a long-term *in situ* conservation of genetic diversity.
2. To promote and facilitate basic and applied research and monitoring.
3. To promote international co-operation.
4. To promote appropriate sustainable management of the living resource.
5. To provide opportunities for education and training.
6. To conserve representative samples of various ecosystems.
7. Top disseminate the experience so as to promote sustainable development else where.

2. *Ex-situ* Conservation

Ex-situ conservation is known as offsite preservation. The conservation of species in suitable locations outside their natural habitat. Such species require protection because its survival in wild may no longer be possible. There are some facilities given for their protection like animal preservation include auaria, game farms, zoos, while plants one maintained in botanical gardens and seed banks.

(*a*) **Botanical Gardens.** The Royal Botanical Gardens of England consist of 250000 species of plants. **IUCN** consider 2,700 of these species endangered or rare. Eight Botanical gardens occur in India. Plant Resources Center at Bhubaneswar, in the world largest botanical garden. The main importance of Botanical gardens is plants supply for research and horticulture. Recently **NBPGR** (National Bureau of Plant Genetic Resources) a special garden is established in New Delhi.

(*b*) **Seed Banks.** These are most important for long terms storage of seeds. Gene banks preserve stocks of both seeds and vegetative plants parts. Seed shows of genetic viability, so its storage ensures conservation of genetic diversity.

(*c*) **Zoological Parks.** Zoo play a vital role in conservation of biodiversity. They give support and re-introduce endangered species.

(*d*) **Aquaria.** The World Conservation Union is develop a breeding programmes for endangered fish species. Aquaria give the captive propagation of threatened fresh water species.

Tissue Culture Technique. Also necessary for conserving "animals cells, ovarian and embryonic tissues etc."

EXERCISES

1. Write defined note on biodiversity and its conservation.
2. Write a note on "biogeographical classification of India".
3. Write short notes on the following:
   (*a*) Biodiversity at global level
   (*b*) Biodiversity at national level
   (*c*) India as a mega-diversity nation
   (*d*) Hot spots of biodiversity.
4. Write a short note on endemic species of India.

❑❑❑

Unit V ENVIRONMENTAL POLLUTION

Pollution can be defined as, the act of process of polluting or the state of being polluted, especially the contamination of soil, water or the atmosphere by the discharge of harmful substances.

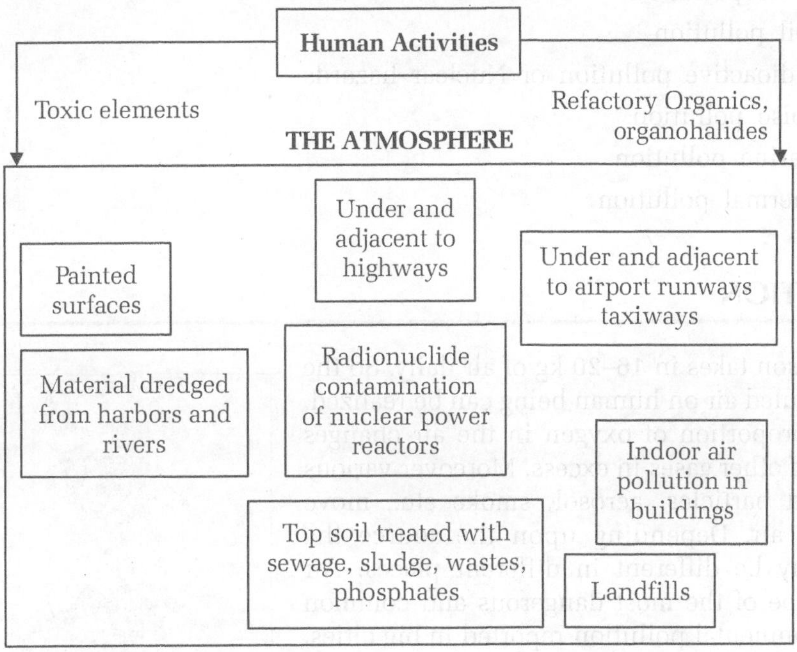

According to **Odum**, Pollution may be defined as, *"An undesirable change in the physical, chemical or biological characteristics of air, water and land that may or will be harmful to human life and other living organisms, living conditions, industrial progress, cultural assets and will deteriorate out raw material resources."*

According to **Southwick** (1976), pollution may be defined as, *"Unfavourable alteration of our environment largely as a result of human activities."*

Classification of Pollutants. Among the different pollutants affecting our environment, two categories may be clearly recognized namely:

1. **Natural pollutants**
2. **Synthetic pollutants**

Of the natural pollutants, carbon dioxide, carbon monoxide, sulphur dioxide, and other trace lancets are the consequence of life processes being produced through respiration, faeces, urine and body decomposition. The monstrous rate of increase in pollution has resulted in a systematic increase of such pollutants.

On the other hand, there is a vast array of synthetic pollutants which are continuously increasing with urbanization and industrial growth. They include pesticides, detergents, pharmaceuticals, cosmetic products, organic acids, aerosols, and metals etc. Several of these compounds are extremely stable and persist in the environment of a considerable period posing serious environmental hazards. In simplest terms, pollution is classified according to the habitat *i.e.*, (air, water, soil), or source *i.e.*, domestic, industrial, radioactive or according to the type of the pollutant viz. gaseous pollutants, inorganic pollutants, organic pollutants, heavy metals, pesticides and detergents etc.

There are seven types of pollution:

1. Atmospheric or air pollution
2. Water pollution
3. Soil pollution
4. Radioactive pollution or Nuclear hazards
5. Noise pollution
6. Marine pollution
7. Thermal pollution

AIR POLLUTION

A healthy person takes in 16–20 kg of air daily. So the impact of polluted air on human being can be realized. The natural proportion of oxygen in the air changes with release of other gases in excess. Moreover various types of solid particles, aerosol, smoke etc., move freely in the air. Depending upon the source the pollutants may be different in different places. Air pollution is one of the most dangerous and common kind of environmental pollution reported in big cities, particularly in industrial cities. The industries, by the way of burning fuels of various kinds release such pollution are:

(*i*) *Automobiles;* such as cars, scooters, locomotives.

(*ii*) *Industrial Processors;* such as chemicals plants, paper mills, cement factories, petroleum refineries, cotton mills, rubber factories, plastic factories, buring kilns etc.

(*iii*) *Heating Plants;* such as thermal power plants, atomic power station, domestic burning of coal etc.

Fig. 5.1. *Industrial Emission.*

Material Sources of Air Pollution

Among the various sources of air pollution the following may be mentioned as important.

(a) Gases

1. **Carbon Monoxide.** It is released by automobiles and factories where coal is used as fuel. Its excess in the atmosphere causes serious complication in man affecting nerves, heart and lungs. It combines with blood hemoglobin to produce carboxyl hemoglobin which reduces oxygen carrying capacity of the R.B.C. It also increases the blood clotting tendency and the rate of fat accumulation in the body.
2. **Sulphur Compounds.** These are released due to biological decomposition, volcanic eruption and from factories using coal and petroleum, geothermal plants, from petroleum refineries. Smelting of sulphur ores also release sulphur dioxide.

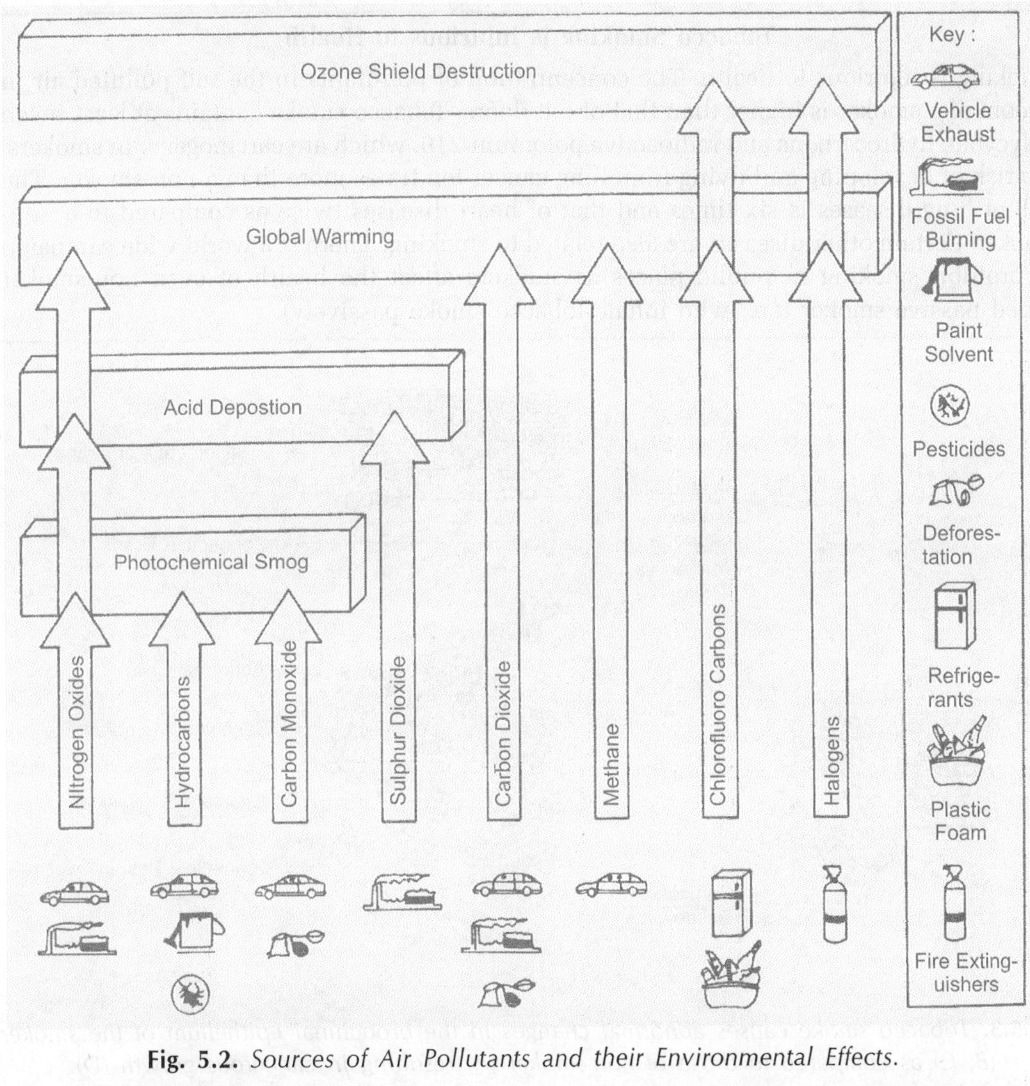

Fig. 5.2. *Sources of Air Pollutants and their Environmental Effects.*

3. **Nitrogen Oxide.** Its main sources are automobile exhaust, power generator, anaerobic decomposition of organic waste. It causes eye and nose irritation and

pulmonary discomfort. It affects plants which is evident by fruit drop and leaf drop.

4. **Hydrocarbon.** Its important sources are automobile exhaust, spill and seepage of natural gas and oil, incomplete composition of fuel, explosives, petrochemical industries and biological decomposition of organic matter etc.
5. **Photochemicals.** Certain pollutants mix with other pollutants under the influence of U.V. radiation and form some compounds such as peroxyacetyle nitrate (PAN), ozone (O_3) and particulates all forming photochemicals smog.
6. **Smoking.** Tobacco smoking in closed chamber, buses and trains affects others and therefore it falls under pollution. Tobacco smoke contains 300 chemicals compounds of which nicotine and tar dangerously affect the health of those who smoke regularly or are exposed to such substance. These are carcinogens.

Tobacco Smoking is Injurious to Health

Smoking is injurious to health. The concentration of pollutants in the self polluted air in a room of a smoker is higher than that of out doors. Tobacco smoke contains at least seven polycyclic hydrocarbons and radioactive polonium-210, which are carcinogens. In smokers, the risk of developing and dying from lung cancer ten times more than a non-smoker. The risk of lung diseases is six times and that of heart diseases twice as compared to a non-smoker. Certain other diseases are also related to smoking. There is a world wide campaign to prohibit smoking in public places as smokers affect the health of even non-smoker called passive smoker (*i.e.*, who inhale tobacco smoke passively).

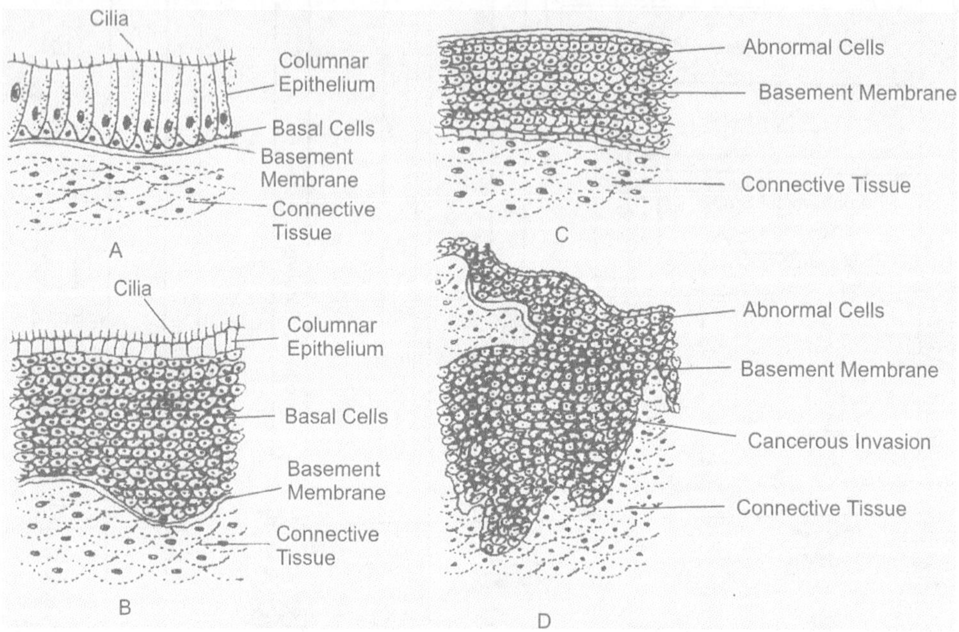

Fig. 5.3. *Tobacco smoke causes abnormal changes in the bronchinal epithelium of the smokers (B, C) as compared to those of non-smoker (A) resulting in cancerous growth (D).*

(b) Solid Particles as Pollutants

1. **Lead.** Lead pollution is more evident in urban and industrial areas than in rural areas. Automobiles exhaust is the main source of lead pollution. Lead pollution is also caused from painting, ceramics, pesticides etc. A over exposure to lead interferes with the development of R.B.C., disrupt functioning of cells, circulatory and nervous system.
2. **Cement and other Dust.** Cement particles affect plant growth in that the leaves fall premature and produce infertile seeds. Coal dust and asbestos cause necrosis in plants. Asbestos may cause lung cancer in man. Dust from thermal power station seriously impairs plant growth because of the presence of pulverized fly as (PFA). Domestic and agriculture dust, pollengrains, microbial spores cause allergy and other complications in man.
3. **Fluorides, Salts and Agricultural Chemicals.** Fluoride when released in excess settles down on vegetation. It causes toxic effect on animals and calcification of their nose and teeth. Potassium salts and sodium chloride also affect plants in various ways.

Effect of Air Pollution on Man and Animals

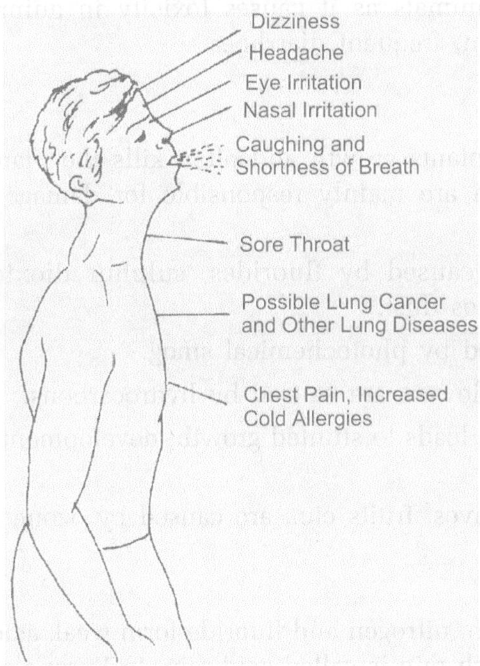

Health Hazards Caused by Air Pollution

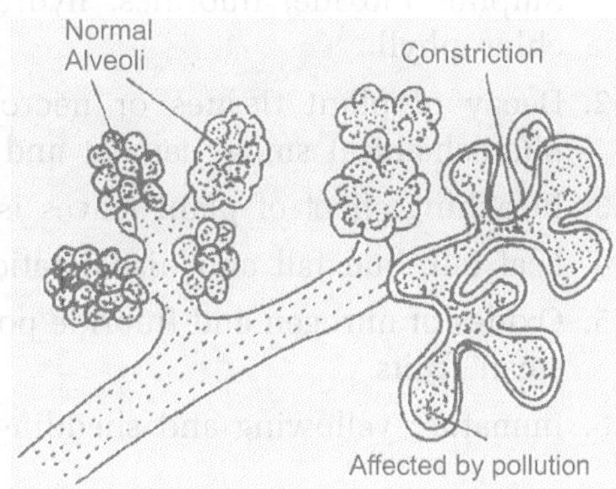

Effect of Polluted Air on Bronchial Tubes

Fig. 5.4.

Air pollutants affect man as well as animals. The reactions in man have been studied. The effect on animals may also be the same. Some of the effects of air pollutants are mentioned below:

1. Sulphur dioxide causes respiratory ailment, headache, eye irritation, irritation of amicus membrane, causing dryness of mouth and throat.
2. Hydrocarbons affect mucus membrane, bronchial tract, lungs. These may cause lung cancer.
3. Oxides of carbon, nitrogen, sulphur cause irritation of eye and nose, affect pulmonary tract. Often these pollutants diffuse into the blood stream and combine with hemoglobin causing reduction in oxygen transport capacity of the blood.
4. Photochemical smog causes severe complications like coughing, headache, pulmonary congestion.
5. Pollen grains and fungal spores cause allergy.
6. Dust and cotton fibers cause pulmonary trouble asthma.
7. Mine dust causes black lung disease. Asbestos fibers cause serious ailment called asbestosis.
8. Smoke affects lungs.
9. Lead pollution from automobile exhaust, pinging ceramics etc., affect R.B.C. circulation of blood and nerves.
10. Fluorides, particularly affect grazing animals as it causes toxicity in animals which leads to lameness, loss of weight, frequent diarrhoea.

Effect of Air Pollutants on Vegetation

1. Loss of chlorophyll (chlorosis) reduces plants growth and often kills the plants. Sulphur dioxide, fluorides, hydrocarbon are mainly responsible for damage to chlorophyll.
2. Decay of plant tissues or necrosis is caused by fluorides, sulphur dioxide, photochemical smog, cement and asbestos dust.
3. Bleaching effect of plant leaves is caused by photochemical smog.
4. Leaf fall, bud fall and discoloration of flowers are caused by hydrocarbons.
5. Oxides of nitrogen and fluoride pollution leads to stunted growth, development of small fruits.
6. Immature yellowing and shedding of leaves, fruits etc., are caused by ozone.

Effect of Air Pollution on Materials

In presence of moisture, oxides of sulphur, carbon, nitrogen and fluoride form weak acids. These acids are brought down by rain water. Such rain is called acid rain. In Europe and in many cities in India such acid rain effect is increasingly noticed. This rain is corrosive to metals, stone, concrete and many other materials. As a result of this, buildings and monuments in many countries including India are badly affected. The famous Taj Mahal of Agra which is near to Mathura petroleum refinery is under severe threat of the effect of acid rain.

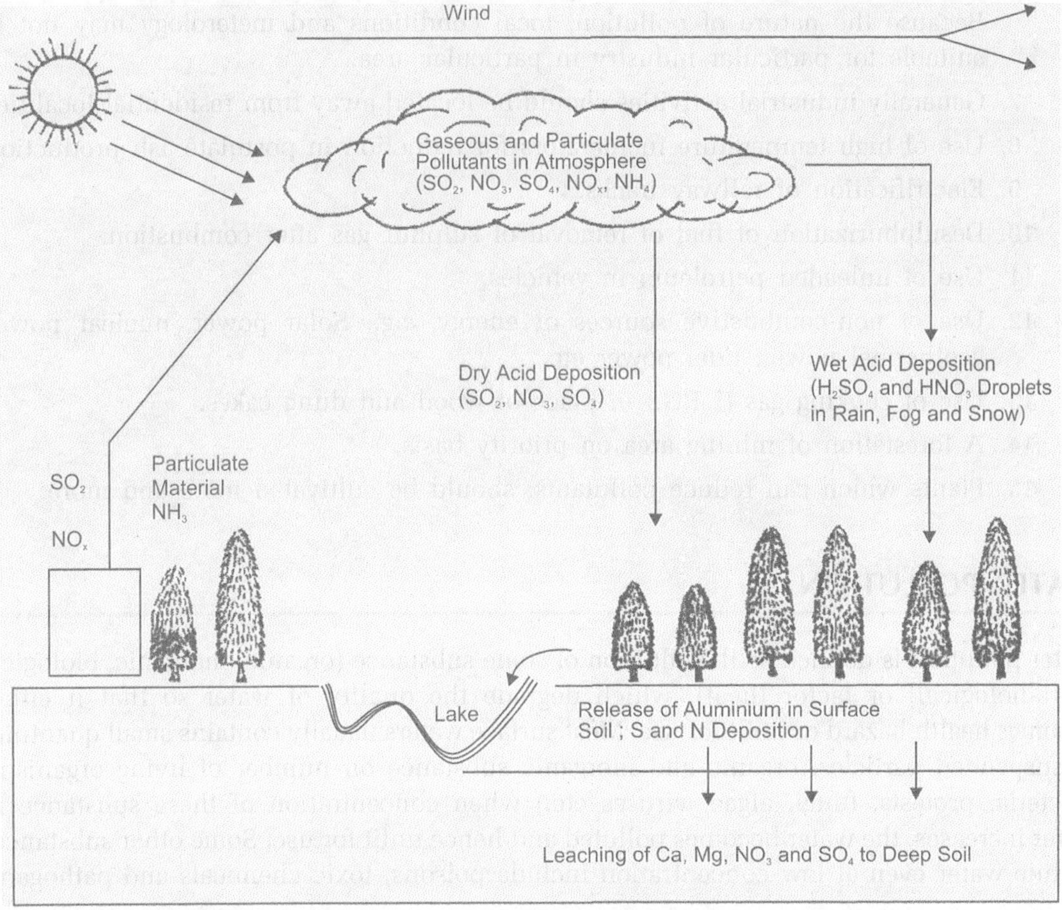

Fig. 5.5. *Dry and Wet Acid Deposition or Acid rain.*

Control of Air Pollution

With a view to controlling air pollution, at first, harmless standard of air quality must be established and then adequate law should be enforced for maintenance of this standard by various industries. Already such laws exist but these are not strictly enforced and the industries are reluctant to follow the norms because of cost involved. For effective control of air pollution caused by industries the following measure may be taken:

1. Use of settling chambers, bag filter, electrostatic precipitator, gas absorber, catalytic combustion agents. Etc.
2. Chimneys should be raised so that smoke and particulate matters are thrown high and are carried by wind for dispersal.
3. Recycling of harmful chemicals and their reuse.
4. Control over use of fuel.
5. Exhaust emission by automobiles much be reduced by mechanical means.

6. Environment engineers should say before setting up industries in particular place. Because the nature of pollution, local conditions and meterology may not be suitable for particular industry in particular area.
7. Generally industrial activities should be located away from residential localities.
8. Use of high temperature incinerators for reduction in postulate ash production.
9. Electrification of railway tracks.
10. Desulphurization of fuel of removal of sulphur gas after combustion.
11. Use of unleaded petroleum in vehicles.
12. Use of non-combustive sources of energy, *e.g.,* Solar power, nuclear power, geothermal power, tidal power etc.
13. Use of cooking gas (L.P.G.) in place of wood and dung cakes.
14. A forestation of mining area on priority basis.
15. Plants which can reduce pollutants; should be cultivated more and more.

WATER POLLUTION

Water pollution is defined as the addition of some substance (organic, inorganic, biological or radiological) or factor (heat), which degrade the quality of water so that it either becomes health hazard or unfit for use. Most surface waters usually contains small quantities of suspended particles, organic and inorganic substance on number of living organisms (bacteria, protests, fungi, algae, viruses etc.) when concentration of these substance in water increases, the water becomes polluted and hence unfit for use. Some other substances pollute water even at low concentration include poisons, toxic chemicals and pathogens.

Sources of Water Pollution

The main sources of water pollution are as follows:

1. **Community Waste Water.** Community waste water include discharges from houses, commercial and industrial establishments connected to public sewerage system. The sewage contain human and animal excreta, food residues, cleaning agents, detergents and other wastes. It is always rich in organic matter, bacterial and other biological pollutants. It is characterized by its possibility. Possibility is the process of decomposition of organic matter present in water by micro-organisms using oxygen. With increase in the amount of organic wastes in water, the bacterias multiply rapidly and use up the available oxygen. Lack of oxygen kills fishes and other aquatic life. Deoxygenating of water causes anaerobic bacterial to produce foul smelling gases. The waste water from industries, when discharged into public sewers, it becomes the source of large quantities of suspended solids and become the source of several chemical pollutants such as heavy metal's cyanides, nitrates, phosphates and ammonical nitrogen.

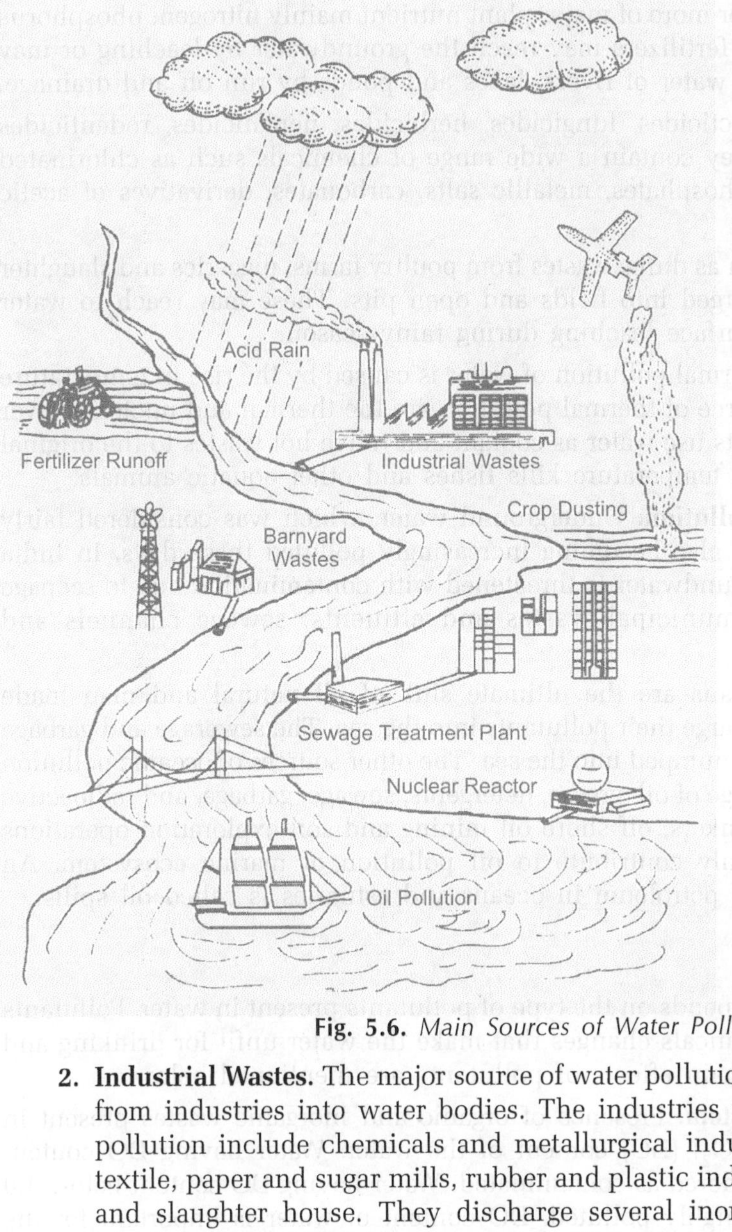

SOURCES OF WATER POLLUTION

Oxygen demanding wastes
Biodegradable organic compounds (*e.g.*, sewage, wastes from food processing plants, paper mills and tanneries)

Plant nutrients
Nitrates and phosphates from detergents, fertilizers, and sewage treatment plants

Sediments
Enrichment soil in water due to soil erosion

Thermal discharges
Heated water from power plants

Disease-causing agents
Bacteria and viruses from sewage (*e.g.*, food poisoning and hepatitis)

Synthetic organic compounds
Pesticides, industrial chemical (*e.g.*, PCBs)

Radioactive substances
From nuclear power plants, medical and research facilities, and nuclear weapons testing

Fig. 5.6. *Main Sources of Water Pollution.*

2. **Industrial Wastes.** The major source of water pollution is the waste water discharged from industries into water bodies. The industries which are the source of water pollution include chemicals and metallurgical industries, food processing plants, textile, paper and sugar mills, rubber and plastic industries, oil refineries, tanneries and slaughter house. They discharge several inorganic and organic pollutants, which may prove highly toxic to the living beings.
   (*i*) *Inorganic pollutants.* They include fine particles of different metal's, chlorides, sulphates, cyanides, thiocyanate, oxides of iron, copper, cadmium, mercury and chromium, acids and alkalies.
   (*ii*) *Organic Pollutants.* They include cellulose fibers, carbohydrates, proteins, oils, fats, phenols, naphtha, organic acids, aromatic compounds, antibiotics and several tier patricide organic compounds.
3. **Agricultural Sources.** Artificial fertilizers and pesticides have become indispensable for the present day high yielding varieties of crop plants. Consequently, they have become a potential source of water pollution.

Fertilizers contain one or more of major plant nutrient mainly nitrogen, phosphorus and potassium. Excess fertilizers may reach the groundwater by leaching or may be mixed with surface water of rivers, lakes and ponds by run off and drainage.

Pesticides include insecticides, fungicides, herbicides, nematicides, rodenticides and soil fumigants. They contain a wide range of chemicals such as chlorinated hydrocarbons, organophosphates, metallic salts, carbonates, derivatives of acetic acid etc.

The animal excreta such as dung, wastes from poultry farms, piggeries and slaughter houses, etc. are discharged into fields and open pits. These may reach to water through run off and surface leaching during rainy season.

4. **Thermal Pollution.** Thermal pollution of water is caused by the rise in temperature of water. The main source of thermal pollution are the thermal and nuclear power plants. The power plants use water as coolant and relive hot wastes to the original source. Sudden rise in temperature kills fishes and other aquatic animals.
5. **Underground Water Pollution.** Underground water, which was considered fairly safe source of water is also becoming increasingly polluted these days. In India at many places, the groundwater is threatened with contamination due to seepage from industrial and municipal wastes and effluents, sewage channels and agricultural runoff.
6. **Marine Pollution.** Oceans are the ultimate sink of all natural and man made pollutants. Rivers discharge their pollutants into the sea. The sewerage and garbage of coastal cities are also dumped into the sea. The other sources of oceanic pollution are navigational discharge of oil, grease, detergents, sewage, garbage, and radioactive wastes. Capsized oil tankers, off shore oil mining and soil exploration operations and oil refineries mainly contribute to oil pollution of marine ecosystem. An accidental discharge of petroleum in oceans and estuaries is called oil spills.

Effects of Water Pollution

The effect of water pollution depends on the type of pollutants present in water. Pollutants brings about physical and chemicals changes that make the water unfit for drinking and harmful to aquatic life. The main effects of pollutants are mentioned below :

Effects on Aquatic Ecosystem. Presence of organic and inorganic wastes present in waste decrease the dissolved (O_2) (DO) content of the water. Water having DO content below 8.0 mg L-1 may be considered as contaminated. Water having DO content below 4.0 mg L-=1 is considered to be highly polluted. DO content of water is important for the survival of aquatic organisms. A number of factors like surface turbulence, photosynthetic activity, O_2 consumption by organisms and decomposition of organic matter are the factors which determine the amount of DO present in water.

SOIL POLLUTION

Soil gets polluted by a number of ways. The major kinds of soil pollution are:

1. Acidification,
2. Saliniation and sodification,

3. Agrochemical pollution,
4. Contamination by metalliferous water.

Acidification is caused by several human activities such as excessive use of nitrogen fertilizers, acid deposition and long drainage resulting from industrial pollution.

Soils that accumulate salts are known as saline or sodic soils. Salt affected soils are common feature of arid and semi arid geographic regions. The soil pollutants which effect agricultural production are heavy metals agrichemicals, radioactive materials and petroleum products. Heavy metals which tend to accumulate in soils include cadmium, chromium, copper, mercury, nickel, lead and zinc. Furthermore, improper management of industrial wastes can have profound implication on soil pollution. Soil plays and important role when an industrial waste is deposited in a land fill or placed on the soil to promote degradation or breakdown. This will decrease the quality of materials available for transport into groundwater.

MARINE POLLUTION

Marine pollution is nay activity that changes quality of sea water and prevents its use of the benefit of human beings. Marine pollution can be caused by domestic source, urban sources, industrial sources, agricultural sources, mining activities, shipping operations drilling of petroleum compounds. Oil pollution of the marine environment is a serious problem.

Pollution of the sea is principally caused by effluents; discharged from countries surrounding the oceans. The important pollutants of the sea include amongst other, the heavy metals, halogenated hydrocarbons, carbon dioxide, petroleum, artificial radioactivity and ultimately the litter. A fatal example of marine pollution through heavy metals was recorded by the residents on the shores of Mina Mata Bay in 1953 and Nigata islands in 1965. Several people died and a number of them crippled due to consumption of mercury contaminated fish. Other metals like lead, zinc, iron, cadmium, manganese, copper, nickel, and silver are often found in the sea water and have been found hazardous to marine ecosystems.

NOISE POLLUTION

Noise has been defined as: The sound having undesirable effects on the recipients. The problem of noise pollution has been three distinct components *i.e.*, source, propagation and recipient. The source may be varied in nature. In different words, it is the byproducts of energy conversion. The automobiles and jet engines in atmosphere, velocity of the wind, temperature changes as well as natural obstruction affect the propagation of sound. The problem of propagation can be overcome by adequate insulation. As far as the recipients are concerned, the primary need is to minimize the period of exposure to reduce the noise hazards.

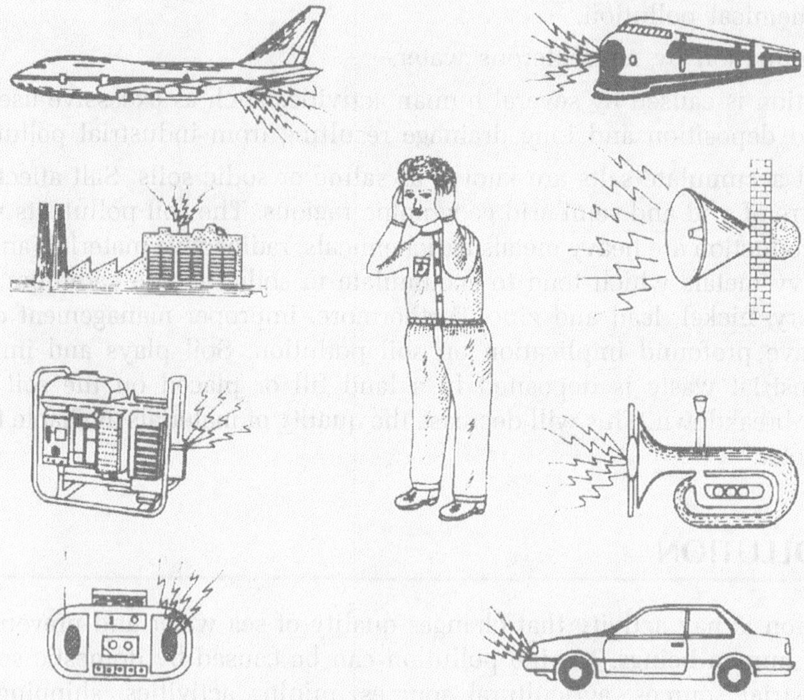

Fig. 5.7. *Chief Sources of Noise Pollution.*

In the terminology of noise pollution, the perceived noise level (PNL) is measured in the units of decibels (DPNL). Indian towns are most noisy than town of western industries. Kolkata, Mumbai and Delhi are regarded to be amongst the most noisy cities in the world. Railway stations, bus stands and airports are the noisiest areas. Nearly 20 to 40 decibels are estimated to be normal while 120 decibels may lead to disastrous effects such as loss of hearing, blood pressure, increased heart beats and tension.

Effects of Noise Pollution on Health. Nearly 20 to 40 decibels of noise are estimated to be normal while 120 decibels may lead to disastrous health hazards. This may include loss of hearing, emotional distress, blood pressure, slipping sickness, increased heart beats and tension. In majority of the developed countries special noise zones are being developed in areas around the airport demarcating areas for development and control.

Noise pollution promotes the development of several non-auditory health effects including speech disturbances, sleep problem, insomnia, annoyance, hypertension and physiological disorder. These are neurosis, anxiety, insomnia, hypertension, sweating, giddiness, nausea and fatigue. Cardiovascular system is specially vulnerable to high levels of noise. Chronic noise may lead to abortions and congenital defects.

Reactions to noise pollution could be listed almost endlessly when one hears a loud noise, the blood vessels constrict, the skin pales, the pupils dilate, the eyelids close, holds the breath and the voluntary and involuntary muscles become tense. Gastric secretion diminishes and blood pressure increase. Adrenaline is suddenly injected into the blood stream which increase tensions, nervousness, irritability and anxiety. Prolonged exposure to such noise can lead to severe mental disorientation and in some cases the violence. According to **Nobel Laureate Robert Koch** *noise is the worst enemy of health.*

THERMAL POLLUTION

Our thermal power plants are endangering life and property by belching out gases at high temperatures such as oxides of sulphur and nitrogen that culminate in acid rains. About 201 million tons of the gas is said to have been thrown out by refineries in 1979. The Mathura refinery alone has become a big threat to historic Taj Mahal, in addition to health hazards likely to occur in heavily populated area around it. Not only the gases emitted by thermal power plants but locomotive engines and cement factories are the chief contributors to thermal pollution. Coal-oil fired generators and atomic energy plants which produce large amount of waste heat to cause thermal pollution. Various industries that require water for the cooling system thrown warm water into the rivers or lakes along with their effluents. This causes warming which produces distinct changes in aquatic biota and may result in ill effects on aquatic ecosystem.

SOLID WASTE POLLUTION

Solid wastes may be classified based partly on contents and partly on moisture and heating value. A typical classification is as follows:

1. **Garbage.** Refers to the participle soiled waste constituents produced during the preparation or storage of meat, wit, vegetables, etc. These wastes have a moisture content of about 70% and a heating value of around 6×10^6 J/kg.
2. **Rubbish.** Refers to non participle solid waste constituents, either combustible or non-combustible. Combustible wastes would include paper, wood, scrap, rubber, leather etc. Non-combustible metals glass ceramics etc. These wastes contain a moisture content of about 25% and the heating value of the waste is around 15×10^6 J/kg.
3. **Pathological wastes.** Dead animal, human waste, and hospital waste constitute pathological waste. The moisture content is 85% and there are 5% non-combustible solid. The heating value is around 2.5×10^6 J/kg.
4. **Industrial wastes.** Chemicals, paints, sand, metal ore processing, fly ash, sewage treatment, sludge, etc., are the examples of industrial waste.
5. **Agriculture wastes.** Farm animals manure, crop residues, etc.

The environmental damage caused by solid waste is mostly uncontrolled. Dumping of urban wastes destroys the beauty of the countryside; also, there is the danger of water pollution when leakage from a refuse dump enters into surface water or groundwater resources. In addition, uncontrolled burning of open dumps can cause air pollution. Solid waste disposal is at present limited to land and the ocean. Some of the wastes can be recovered and reprocessed by a procedure commonly known as recycling. However, before the wastes can be disposed of effectively, they must be collected efficiently. These activities. *i.e.*, collection, disposal and recovery form a part of the solid waste management system. The old 'dilute and asperse' concept of waste management no longer works and newer concept of 'concentrate and contain' may give way to a concept called instead waste management.

Health Effects of Solid Waste Pollution

The relationship between solid wastes and human disease is difficult to prove. Nevertheless, improper handling of solid wastes is a health hazard and cause damage to the environment. The main risks to human health to arise from the breeding of disease vectors, primarily, flies and rats. A common transmission route of bacillary dysentery, amoebic dysentery and diarrhoea disease in India is from human faeces by flies to food or water and thence to humans. It has been estimated that; in warm climates, exposed garbage produces as many as 70,000 flies per 0.03 m^3 in a week.

The refuse dumps are the source of food for rats and small rodents which quickly proliferate and spread to neighboring area. Rats destroy property, infect by direct bite and spread various diseases like a plague, endemic typhus, salmonellas, trichinosis, etc. Apart from diseases for which insects and rats are carrier, the handling and transfer of biological wastes poses a threat to the worker and to those they contact. Disease transmission may occur through direct contact with the wastes, through infection of open sources or through vectors.

The hazardous wastes are injurious to human health; some have acute effects while others pose a health hazard after prolonged period of exposure. Improper disposal of such wastes has resulted in the death of humans and animals through contamination of crops or water supplies.

URBAN AND INDUSTRIAL WASTE

Urban and industrial wastes are responsible for all kinds of pollution that is air, soil and water pollution. Hence it is necessary to know the factors which cause to release their wastes. Until and unless we know such causes, proper controlling system of pollution cannot be managed.

Land Pollution

The one fourth of the total area of the surface of the earth is soily and the rest contains water. Since the time industrialization has started, a great change has affected the soil of this earth. Industrialization has caused the new cities to come in existence and the old cities to swell more and more. Consequently urban life expanded, with a simultaneous result of over population. Naturally, the new habitations and the projection of old little cities in the form of big ones, emerged in the shape of a major factor, responsible for the land pollution and in turn, the water and air pollution.

The surface part the large land, which is bestowed by the nature to man, naturally is under his care and supervision. Through initially, the mankind was free to use it in any irresponsible way, but the present position which has occurred before him is alarming. Inspite of the fact that human nature has not taken an considerable turn to cope with the problem arisen due to his activities. Land, air and water pollution are interrelated to one another as one cause to infuse pollution in the other. This factor is to be taken in serious consideration, to manage the pollution factor.

Degradation of Environment

First of all, we must consider the causes which degrade it to an inevitable extent.

Environment is the cycle arrangement of nature, related to the ecosystem of the climatic regions. Every climatic region has its own specialties, but nature has got its unique speciality, that is, to keep all the ecosystems in such uniformity that these may not be disarrayed. Whenever disintegration of the ecosystem of any region seems tending to be disintegrated from the chain, it changes its shape and form, because there can be no place in this universe without having its own ecosystem, but latently connected with the huge structure of the universal ecosystem. The scientific inventions have proved this beyond doubt that the environment as an embodiment itself is undoubtedly universe factor. Naturally it is affected by the human activities. Nature has bestowed its attributes in abundance upon the human race, as if these all are made for it. But contrary to this it has been felt that human being has not remained kind to nature. His achievements are the result of his unkind exploration of the nature, which has explicitly exposed his starvation for the amenities, power and wealth at the cost of spoiling environment. It is an apparent fact that the activities of human being are significantly responsible for the degradation of environment. In the near past, particularly since the industrial revolution man's hunger for wealth developed up to unending limits. However, the man, in the lake of industrial storm caused the environment degraded to the present declinated condition. In some way or the other, human activities occur without keeping the nature's role in view. Usually land is the first victim. In case of soil erosion, nature's role is very significant. Time factor remains. Activating its process of change very slowly. Human being, the main biotic organism through highly sensitive to changes, not possibility perceived this for a long period of centuries. Contrary to these, human activities usually remained causing the soiled and other components of environment affected.

Besides this course of nature, which is latent, it has got the accidental role also, in the form of volcanic explosions, droughts, floods, landslides, earthquake somehow or the other it is also believed that in certain causes of natural happening *e.g.*, droughts, landslides, floods, etc., human activities may also conclusive ultimately pollution is also the outcome of natural calamities and human activities hence conservations is needed.

ROLE OF AN INDIVIDUAL IN PREVENTION OF POLLUTION

"Apart from the measure usually taken or required to be taken to save the human civilization from the pollution hazards, the most significant thing is this, that the human being is needed to be personally civilized."

An individual, being the member of society is required to observe that he has also to reform himself in order to maintain the purity of environment, generally, the man has got to perform multifarious roles as a member of the society but ignores his duties, he is expected to observe towards it. Though the most important of all is to confine himself to us activities which are necessary to maintain smooth relations with other member of the society. He should realize that there are other people also who are affected by his behaviour and activities. Social\behaviour not only consists of his observation in respect of reciprocal attitudes, but it is he to consider his working and thinking from the point of view, that he is obliged to perform such other duties also, which usually seem to be unconcerned to his self-interest, but are conducive to social welfare and needed to save environment from pollution.

An individual person is required to observe that he, in position of his being related to the society, must not behave or act adversely to the social welfare. Social welfare is directly related to the environment, which is usually affected by the human activities and polluted thereby. The man thus cannot be expected to cause the environment polluted by his activities. But the picture of the man's role in polluting the environment is very contrary to this expectation.

For near about a century since Europe entered into the field of industrial development. Nobody took a little notice of environmental pollution. Because nobody was familiar to the world environment of pollution as a social concept. But at the time people were of the notion, that human being is at all free to do anything. Hence in the long run of industrial civilization the man did everything which was conducive to the pollution of environment. Naturally by this time pollution remained in process and took form of a very conspicuous problem. However since was considered to be an unbeatable weapon by the European's but when scientists came across to it, they found astonishingly it very difficult. They also looked into the irresponsible and pollutive activities of man. They realized, particularly that the class of human society which was enjoying the major benefits of industrialization are developed civilization were the main culprits. Through these were supposed to be civilized, they did never care in respect of pollution, for which they awere themselves responsible. And when it was also realized that this class was not habitually to imagine about the adverse results of their activities, the Governments took assistance of legislation. These circumstances, however made the man conscious to their activities to some extent.

Governments also realized that it should also try to take cooperation of the society to struggle with the problem of pollution.

Hence, it is badly needed that the man must play main role to solve this problem, in the following manner:

1. Man should be careful in respect of his activities, which become the cause of polluting public lanes, streets and roads. He can abide certain public lane rules in cooperated by the municipal organization. He must also be careful in keeping house drain system free from general domestic solid residues which are supposed to be thrown at dumps.
2. Cooking is usual domestic affair, hence the man can avoid cooking by wood or coal fire, by getting it replaced by smokeless hearth in order to save the atmosphere from smoke.
3. A man, who does not have proper sewage facility in this home, must keep the latrine system very intact and covered. Its cleaning process must also be unexposed.
4. Every person should trained his family and children not to be habitual of dirty practice, in respect of domestic and personal waste. If there is some developed system maintained by the municipal body of the city or town, it should not be overridden.
5. So far as the industrial citizen are concerned they should properly abide to the rules and laws of pollution. Furthermore they should also be morally conscious

in respect of their activities which may be injurious to public health and taste. He should also cause to keep in watch slums, where the industrial labour lives.

6. There are open wells and ponds in villages, generally the people use these ponds for animal drinking. Therefore at such places cleaning their clothes and domestic utensils etc., are prohibited. Furthermore the man should not use these water sources as dumping places.
7. An agriculturist should not allow the outflow of water of the fields, whereas pesticides are abundantly used mixed to rivers, lakes or some other water sources used by public generally.
8. The dung should be deposited in compost pits outsided the human inhabitation, for example villages, towns or cities.
9. Proper arrangement should be made for excretion of faeces, in the manner not be a pollutive cause to the atmosphere.
10. On the whole what is significantly needed is the general consciousness of the activities which may cause pollution to the environment so that a man may refrain himself and others from doing them.

In the end we can add that man usually in his individual behaviour does not feel significant to perceive the reactions of his minor activities, which are the parts of his habitual conducts, harmful to the environmental pollution. But when such major activities are taken in notice collectively, the actual position is; uncovered. It is only an individual's consciousness, in this respect, that prepares him to do something cooperative to the tedious efforts, being in vogue to face the environmental problems. But the trouble is that is his habituate conduct he cannot decide the way which can be proper and significant for his involvement in this direction.

Hence as an individual man does not find an easy way to play his proper role in the solution of environmental problems because these vary in nature and in their awful abundance. To work in the field of environmental problems, adjustment of human role is to some extent difficult. It shall be therefore very necessary that a person who is willing to do something in this direction should choose such work, which can be performed by him conveniently and in his full capacity. There are so many institutions who are at worked in this side. Any person can render his services according to their programme.

DISASTER MANAGEMENT

Nature has got its own way to run this universe, of which this earth is nothing more than a particle. The environment of this earth is merely a fine cosmetic arrangement, mystery of which is not known as yet. Though the scientists are trying their best to know all of its hidden mysteries, but they have not yet achieved a remarkable success in predicting about the natural disasters. These occur suddenly and snatch innumerable happily lives, enjoying the wonderful procurements of human attempts. Hence such formidable happening of nature are fiercely horrible and they demand from the modern scientists their prediction on one side so that rescindive measures might be taken rescue from the disastrous condition with immediate effect, so that heavy loss of lives and property that might not incur on the other side.

However till now the science has failed to predict in respect of such type of calamities, hence to save the humanity from this type of predicament, only the agile management can be of any avail.

Happenings like fire, floods, landslides, earthquakes, volcanic explosion, cyclones, tornados, droughts etc., are the enactive incidents, which cause loss of life and damage to property. The magnitude and frequency of these process depend on such factors as climate, geology, vegetation and human activities. At present when science has caused the man successful in the achievements of tremendous power resources, he has himself also become a significant factor responsible for a lot of such mishappening.

To face the above noted processes and hazards the most necessary thing is to evaluate catastrophe.

Catastrophe, has got its own magnitude, one is concerned to the situation, where damages to property, people and society in general are of most serious nature, thereby require a considerable long time arrangement for their recovery and rehabilitation. It is a long involved process. That is why it is said to be of high catastrophic potential, while coastal erosion, frost, lightning and expansive soils etc. are of low catastrophic potential.

United Nations, have tried to give a rough estimate of the damage caused by natural disasters in between the years 1909 and 1976.

Over 2.6 million death were claimed by earthquakes alone during this period. Floods are the number two claimants for deaths. During this period 1287645 people died and 17220220 were affected becoming homeless and injured. Cyclone is the next to these disaster, which devoured 400000 people.

Earthquakes

Earthquake is a rapid and discernible tremor or movement caused by fracturing of rocks in the earth's crust. Earth movement are the results of large and internal forces of earth movement. These forces are contraction, expansion, uplifting, dearrangement, folding and faulting, which causes earth movements. Indian soil has also become an unwarned victim of such type of earth movements and suffered heavy losses. In 1885, **Oldham** prepared a list of earthquakes, that hit India in historic period.

Reservoir induced earthquakes are the result of human activities because radical water reservoirs are constructed by man. Koyna earthquake of Dec. 10, 1967 was the products of water reservoir which claimed 200 human lives and injured more than 1500 persons along with the disastrous situation arisen in the form of thousand of inhabitants becoming homeless.

Flood

Relatively high flow or stage in river marked higher than the usual also the inundation of low hence which may seem to therefore can be defined as a flood. It is actually a physical phenomenon. Flood, can be caused by abnormally heavy precipitation, dam failures, abundant snow melts, burst weather blockades of rivers. Flood are also the cause of disastrous destruction. The total number of people affected worldwide comes next to those suffered from droughts. Floods include three kinds of floods; Flash floods, River floods and Coastal Floods.

Flash floods are usually caused by towering cumulus clouds, severe thunder storms, tropical cyclones. These also occur during the passage of cold weather forests.

River floods are generally caused by heavy rains, melting of the intense accumulation of snow. Precipitation over large attachment areas is also one of the significant causes of river floods.

Coastal floods are generally the results of tropical cyclones, hurricanes and typhoons. Sometimes rain water, when aggravated by wind induced storm rung along the coast. Sometimes salt water may also be the cause of flood, particualy when in combination of effects from high tides, strom snugs or tsunamis. Recently tsunami waves have caused havoc of most dredful nature, which might have never before seen by the world.

Flood Control

The measures, which can be taken to control the floods are as follows :

1. **Channel Improvements.** If the river channels are improved by getting their beds deepended and widened, flood strength can be controlled. Naturally, when the capacity of river channels is increased flood area can be reduced.
2. **Construction of Relief Channel.** Relief channel means the channel constructed to divert the waterflow from the main channel to other direction. This is feasible method, easy and not as costly as can be the displacement of any coastal settlement from its original place.
3. **Dikes and Dams.** These are retention facilities which can be useful in the cause of flood control. Construction of dikes and dams may retain the floods level watger, which can be easily released under proper management.
4. **Flood Proofing.** Public may be helpful by adopting the method of getting their buildings strengthned to procure safety from flood currents. Naturally, if the foundation of the buildings are laid down on susceptible tolerate the flood strength shall not be dangerous.
5. **Protection Against Erosion.** This can be done by causing the stream beds stabilized with stone, masonry or vegetation. This method is significantly useful, when adopted specially near the bridges.
6. **Site Improvement.** Proper elevation of sites can be an effective method for dwelling place.

Drought

The US weather bureau defines drought as “a period of dry weather of sufficient length and severity to cause at least partical crop failure”. Generally drought is related to crop failure which is due to the reduction of crops are grown in a climatic cycle. Hence the period, in which the climatic cycle fails to provide expected water or moisture in any area, drought occurs there. Technically water scarcity below about 80% of the average availability over the preceding 20 years might qualify as a drought.

Kinds of Drought

These are as noted below:

1. **Meterological Drought.** A situation when occurs due to about 25% decrease from normal precipitation over an area is called as meteorological drought.

2. **Hydrological Drought.** Prolonged meteorological drought, when it takes such a long time that surfacial water is caused to depletion and the result comes in the form of drying up of reservoirs, lakes, ponds and rivers, it is called hydrological drought.
3. **Agricultural Drought.** This is caused by inadequate moisture and rainfall during the healthy crop season. In other words when the major rainfall and moisture becomes un-suprortive to the crop to mature and thereby a heavy crop faliure is results, it called agricultural drought.

Drought Management

There are so many measures suggested to counter the drought conditions in the form of their mitigation and reduction. These can be mentioned as below:

1. Getting seeds in place, when the rain begins.
2. Improvement of range land management.
3. Improvement of water resource management by:
   (*a*) Digging new wells.
   (*b*) Improving the quality of existing wells.
   (*c*) Construction of retention dams or catch dams.
   (*d*) Improvement of catchments, which trap water.
   (*e*) Construction of surface dams or trapping water in sandy river beds.
4. Planting of drought resistant crops.
5. Implementation of counter desertification measures such as.
   (*a*) Erosion control
   (*b*) Tree planting
   (*c*) Construction of wind breaks and barriers.

Cyclones and Anticyclone

These are large rotary wind systems which move across most of the earth areas outside the equatorial belt. Usually the belts of clouds, rain and snow are the factors which cause cyclones, and in contrary to these factors anticyclones, which have lighter winds are generally free from prediction. The diameter of these wind systems is generally comprised of near about 1000 km to 4000 km. The cyclone winds rotate in anticlockwise direction while anticyclone rotate in the clockwise direction. Their rotational nature in the Northern and Southern hemispheres in contrary to each other. In Northern Hemisphere the direction of cyclones and anticyclones are opposite to those in Southern hemisphere.

Origin of a Tropical Cyclone: Hereinafter, the prerequisites of a tropical cyclone are mentioned.

A tropical cyclone is likely to occur when prerequisites are there to give its birth as :

(*a*) Sufficiently high, that is, altitude 5.6 degree for the coriolis effect is desirable. Coriolis effect means the effect of earth's rotation on moving bodies including air current.

(*b*) About 27°C temperature warm water surface is also a requisite factor, because it supplies the overlying air with large amounts of water vapour.

(*c*) Pronounced instability which exists in the air columns or relatively low pressure at the surface. This often causes an anticyclone aloft.

(*d*) Non-presence or a little prevailing vertical wind shear.

Management of Natural Disasters

Since the scientific development so many measures have been taken to know the apprehensive turns atmosphere, which might be culminate to a natural disaster. For this purpose radar units are installed at suitable points. Radar system is capable to warn against storms within a vogue.

A close watch of the clouds, can give the pre-knowledge of cyclonic storms. The spiral bending of dense clouds and rains is the prominent characteristic of cyclonic storm.

Satellites transmit photographs of any points of the cloud system and conditions, as such reveal attitude of the wind movement which is very well efficient to aware of the environmental conditions around the earth. From a considerable remote areas where no other method can prevail satellites system penetrates.

The presence of the legislation has also caused the rules laid down in respect of construction of high building, in order to make these secured from hurricanes.

Generally, as it is possible to secure the man from natural disasters, there are two kinds of measures, taken by management 1. Mitigation and 2. Preparedness. Both of the activities are focused on the following points:

1. As it is found that mismanagement is the main cause behind the most of the environmental degradation and natural disasters, it should be made tight and vigiled to the extent of human efficiency in this respect. Linkages between the environment, natural disasters and development are needed to be specifically established to mitigate disastrous and development of environment.
2. Disaster relief must not be the main aim in place of disaster mitigation. Vulnerable areas should be linked with the development projects so that these might be fully explored for disaster mitigation. Development project are required to be so formulated that these might be able to minimize their adversities.
3. Adequate attention should be given to the economic impact of natural disaster. Cost benefit measures must be taken in course of prime consideration and as such on the cost analysis basis probable disaster events should be evaluated from the mitigation points of view.
4. Area specific, hazard specific advance mitigation plans must be prepared by the government. As it is the national cause, the state government and the central government must not be allergic in providing cooperation to each other. Adequate resource should be managed to be ready to mitigate the environmental hazards. Along with this strong administative set up is the most needed measure.
5. National Center for Information Research and Training for Disaster Management should be established for the training and research in the sphere of disaster management.

ENVIRONMENTAL DISASTERS CAUSED BY POLLUTION CASE STUDIES

Scientific advisory committee to the US President has observed that pollutants have altered the situation of environment of global scale. Pollution has become the cause of massive mortalities, damages and destruction.

Accidental release of toxic, inorganic chemicals and chlorine have caused so many disasters in this world. History witnesses these disasters today known in the form of episodes, such as London Smog in 1952. Meyug River Valley episode in 1930, pollution caused by zinc smelter, steel mill and sulphuric acid plant seized 40% of population of Donora (Pensylvania) in 1948.

London Smog again in 1952 spread over its fierce claws over the city and hooked nearly thousand elderly and sick people. Similarly Eastern and United States, Western Europe and Japan also witnessed the similar scenes and thus air pollution become a world wide episode. Bhopal in India also faced it in 1984.

London Smog

In 1952 Heavy Smog disaster appeared in London where high sulphur dioxide (1.3 parts per million) content caused to create heavy smong (smoke and fog) conditions (December 5.9, 1952) and killed 4000 people in one week. In the year 1956 (December) London faced one more smog episode which took two lives, then in 1956 again smog giant asked near about 700 to 800 lives. In 1962 smog in same city claimed 700 lives.

Smog disasters are not new for London. Record discloses that it was Feb. 1850 when London faced the first smog disaster which claimed thousand lives. The following table shows the smog episode.

Table 5.1. Smog Pollution

Date	*Place*	*No. of Coast Line*
Feb. 1880	London	1000
Dec. 1930	Meyug Valley (Belgium)	63
Oct. 1948	Donora (Denusy Luvia)	20
Nov. 1950	Pozurik (Mixico)	22
Dec. 1952	London	4000
Nov. 1953	New York	220
Dec. 1956	London	1000
Dec. 1957	London	700-800
Jan. 1963	New York	300
Nov. 1960	New York	168

Mediterreanean: Dead Sea

This sea has coastline of 48 thousand kilometer whereabout hundred million people have their habitat. It is surrounded by Europe and Gibraltar and a straight connects it to Atlantic Ocean. There are situated so many luxury tourist hotels. Some industrial establishment are also located. The industries dump their waste into the Mediterranean Sea, which also receive the sewage from the Europian countries. This conditions has highly polluted this area. Lead, Mercury, Chromium etc. in heavy quantity are the components of this refuse. The Pollution caused by the residue of sewage is so severe that emits offensive odour in high density. This state has become a usual affair of closure of the tourist hotels for about six months every year. Since 1950, it has been declared a Dead Sea, unable to support any aquatic life (fish, plants etc.).

Nuclear Explosion

The two atom bombs were dropped by USA during World War II (August 6 and 9, 1945) on Hiroshima, Nagasaki in Japan. These instantly killed six lacs of people, destroyed the two cities and unleashed radioactive fall out which has caused generation to suffer including genetic disorder. Radioactive pollution still continues to damage plants, soil and biosphere in the region.

Minamata Disease

At Minamata bay of Japan (1953–1960) about hundred people lost their lives and thousand were paralyzed after eating mercury contaminated fish from the bay. There is a chemical company named Minamata Chemical Company, which released its waste in the bay. These waste contain Mercury, Sea fish thereby are forced to pick up mercury which is concentrated in their bodies as the Methyl derivative. This was highly toxic. Thus when the Japanese who consumed such fish, ultimately become the victim to mercury. This type of mercury poisioning is known to be Minamata disease.

Bhopal Disasters (Dec 3, 1984)

Leakage mishappening of Bhopal in an american industry named Union Carbide Corporation in the midnight of 2nd and 3 Dec. 1984. In the midnight of 2 and 3 Dec. negligence on the part of Union Carbide Corpn. Bhopal become the cause of taking more than ten thousand lives of Bhopal City, where it existed. This is the worst disaster in human history. It was pesticide factory. There a large volume of Methyl Iso Cyanate (raw material for production of the pesticidecarbon) was stored in the factory. Which leaked into the atmosphere of Bhopal on Dec 3, 1984 at around midnight. Very soon the city was transformed into a gas chamber. Within week ten thousand people died. Thousand people turned blind and lacs of people continue to suffer from various disease. This was mass murder in recent history and the victim survivors are yet to receive their compensation from this corporation or the Government.

It was the end result of negligence on the parts of the central and state government and factory management on one hand and lack of awareness among the public and hospital doctors on the other hand. The doctors become perplexed, stunned and extremely nervous, when the thousand of dead and half alive bodies found into the hospitals.

Chernobyl Disaster 1986

The worst, nuclear reactor accident occurred at Chernobyl USSR (Now CIS) on 28 April 1986. The reactor exploded as a result of uncontrolled nuclear reactions, radioactive fuel and debris shot up into air like a volcanic explosion and spread out in the surrounding areas. The accident killed about 2000 people and damaged soil, water and vegetation in an area of 60 sq.km. around Chernobyl. Several generations in the region suffer from radiation induced diseases including cancer.

Gulf War Hazards (1991)

The Gulf War took place in the year 1991 USA led multination coalition force under the banner of United Nations attacked Iraq to get Kuwait Liberated from the clutches of Iraq in violation of international law and order.

Iraq and Kuwait both were caused to be severely suffered and damaged. Naturally many cities of these two states were destroyed and their environment was polluted to such as extent that it could only be recovered in a long periods of several decades.

About one lac bombs were dropped in Iraq. Naturally Iraq was covered by the cloud of black smoke and dust. About 700 wells were ablaze in Kuwait for over 10 months Black smoke and soot were shot up at high altitudes of the atmosphere. These moved with winds and encircled the earth, warmed up the upper atmosphere. The result was the temperature inversion (Warm air over cool air mass) and the disrupted; monsoon circulation in the regional din acid rain, crop damage in South East Asia. The global temperature dropped by about 0.5 C total of 200 million gallons of oil was dumped into the Persian gulf which destroyed the marine ecosystem.

The end of war displayed the scene Hell on Earth chaos and oil fire, smoke, oil spill and a landscape of mines. This was terrible environmental disaster. To handle this was not easy. It was most challenging. However it was handled by UNEPO (United Nations Environmental Programme). In which so many other countries also cooperated. But except the destruction environmental loss could not be replaced.

Definitely it was a slaughter game of Western countries, which can get us remembered such games played by the culture of Middle Ages throughout the then known world. Though it has not been termed as invasion but in continuance of this incident, after a lapse of short time America and Britain attacked Iraq in the style of an invasion which scahsesd the same type of disaster again to be witnessed by the world. Again, the political affairs and are arising in the same fashion, and USA is nearly prepared to invade Iran. However there might be reasonable causes from the point of view of restricting the irresponsible countries from becoming atomic power, but there are so other circumstantial evolvements which cause it to seem that political measure have become nearly exhausted and the world is proceeding towards the III world war which shall have the world to a damaged environment.

EXERCISES

1. What is the meaning of pollution ? Write main types of pollution.
2. What is the meaning of air pollution ? Write a short note on air pollution.
3. What is the meaning of soil pollution ? Write its main causes and effects.
4. What is the meaning of ocean pollution ? Write its main causes and effects.
5. What is the meaning of noise pollution ? Write main causes of noise pollution.
6. What is the meaning of water pollution ? Write important causes and effects of water pollution.
7. What is the meaning of thermal pollution ? Write its causes and effects ?

❑❑❑

Unit VI SOCIAL ISSUES AND THE ENVIRONMENT FROM UNSUSTAINABLE TO SUSTAINABLE DEVELOPMENT

Many countries of the world are facing social, economic and environmental crises today. Policy-makers find it difficult to formulate plans than can work under present conditions of technological materialism, economic over exploitation and diminishing resources. Contrary to neo-classical theory of continuous material growth, uncontrolled economic activities can directly undermine the potential for development by over-exploiting natural resources and drastically reducing the carrying capacity of the natural ecosystems. Remains of the past civilizations, offer testimony to this concept. These civilizations were not able to cope with the pressures of the degraded environment. Kautilya, the wise ministers of Chandragupta Maurya said that *"stability of an empire depended on the stability of its environment."* In other words, he emphasized the concept of sustainable development. Recently Agenda-21 emphasized the same concept *i.e.,* a deeper understanding of the interconnections between water, nutrients and the biogenic cycle and the energy flows on land, in oceans and in the atmosphere, is crucial for evolving sustainable development strategies.

Mahatma Gandhi referred to a satisfactory society, long before we ever perceived the present meaning of environment, by saying that, *"earth can provide for all the needs of humans, but not for their greeds."* Despite the fact that the terms sustainable development has become common currency among many groups, it is a confused and sometimes contradictory idea and there is no widespread agreement as to how it should worked in practice (Doves and Handmer, 1993). The concept has developed in the international forum form a document called the **World Conservation Strategy** (IUCN/UNEP/WWF, 1980), which argued that three priorities should be incorporated into all development programmes :

- Maintenance of ecological process
- Sustainable use of resources
- Maintenance of genetic diversity.

Sustainable development become popular after the World Commission of Environment and Development (also known as the Braundtland Commission after its chair Gro Harlem Brundtland of Norway) which was formed by the UN in 1983 and reported in 1987 (**WCED**, 1987). The Commission emphasized that the integration of economic and ecological systems is all important if sustainable development is to be achieved, and coined a broad definition of sustainable development which is often quoted, *"development which meets the needs of the present without compromising the ability of future generations to meet their own needs (***WCED***, 1987)."* Subsequently, similar calls for sustainable development have been made by various intentions groupings (*e.g.*, **IUCN/UNEP/WWF**, 1991), notably at the UN Conference on Environment and Development (**UNCED**), otherwise known as the Earth Summit, in Rio de Janeiro in 1992.

URBAN PROBLEMS RELATED TO ENERGY

Today we need energy for agriculture, industry, transport, communication, comfort and defiance. Energy plays a vital role in the socio-economic development of a society. Presently, no one can think of a comfortable life without energy. As compared to rural areas, the energy requirement is very high in urban areas. Regular supply of electricity has become indispensable in high rising building, multiplexes and shopping malls of the cities. The activities come to stand still in such establishment, when there is an interruption in power supply. Therefore, most of such establishment have their own very high capacity generator sets for power back up. These generator sets cause hazardous air and noise pollution.

In many cities and towns, water supply is dependent on electricity. Acute shortage of drinking water is faced by the people living in these areas during the peak summers due to short supply of electricity. These days, many of the house jobs such as cooling of food, washing of clothes, cleaning of houses etc. have become electricity based in urban areas. Shortage of electricity make the life miserable, of the people living in cities.

Human population uses energy in disproportionate manner. For example, 5% of the world's population use 25% of the total energy consumed in the world. The use of wood in USA was hottest in 1880s, when the use of coal was widespread. The use of coal began to decline in 1950s, when oil and gas started to become available. The fossil fuels were used to generate electricity. Electricity is the most useful form of energy. The production of electricity and centralized distribution systems have resulted into economic growth and development of 20th century life-styles in developed countries.

While energy has brought modern economic development, it has also resulted into world's most significant environmental problem *viz.* Photochemical smog in Las Angels and Mexico city, acid rain, air pollution, global warming, oil spill and loss of fish and wildlife. Further economic development will require more energy. Therefore, energy is to be conserved.

WATER CONSERVATION

Water is a basic need for our daily life. Water conservation is essential for economical growth. The African countries are facing a major water scarcity problem. The latest United Nations reports give information about the depleting potable water supply. About 60% population live under severe water stress. Several types of dams constructed for water storing purpose. Water is a vital natural resource and its quality and quantity conservation are necessary. Today there is a massive growth of population leads to water crisis. Some improper acts generate a water problem like exploitation of groundwater, without maintenance surface water systems etc. There are some strategies accepted for the conservation and utilization of water on sustainable basis.

Industrial Conservation. Cooling towers used for less water utilization. For treating the effluent industries and manufacturing units should install. Recycling method used for this purpose. Recycled water used for floor washing.

Domestic Conservation. Domestic sector uses a large amount of water. So its conservation is also necessary and use water in proper way. Less quantity of water use for bathing purpose. Check the leakage and use less amount of water for toilet flushing.

Agricultural Conservation. Agricultural sector requires water for irrigation purpose. Wastage of water occurs due to leaks in irrigation pipes, barren areas require water and run off. For reducing it we use some conservation measure:

1. Use the sprinkle irrigation.
2. Some techniques like tillage, inter-cropping, ground cover on drainage ways may be adopted.
3. Minimize seepage and evaporation.
4. Using mulching for excess water removing.
5. Watering the plants in the morning or at night because evaporation is minimal.

There are some other strategies that can support water conservation activities :

Sustainable Water Utilization. (1) Minimum domestic water consumption, (2) Improve irrigation losses, (3) Maintain, (4) Thermal and electrical energy economy identified.

Water Conservation. Water is basic need for our daily life. Water conservation is essential for economical growth. The Africans countries are facing major water scarcity problem. The latest United Nation reports give information about the depleting potable water supply. About 60% population live under severe water stress.

Several types of dams constructed for water storing purpose.

RAINWATER HARVESTING

Rainwater harvesting process promotes conservation of water. This water use for livestock, landscape, agricultural and household. There are some other source of water like well drilling equipment, development of large, reliable water treatment and distribution systems. etc. Rainwater is scum. Rainwater harvesting method use for collecting, inducing, conserving and storing local surface runoff for further use. Rainwater harvesting is an essential for us because surface water is inadequate to fullfils our demand. Its need is increasing due to following reasons:

1. Shortage of water in arid and semi arid regions.
2. Rainwater has beneficials for health.
3. Potential cost saving related to rain water collection systems.

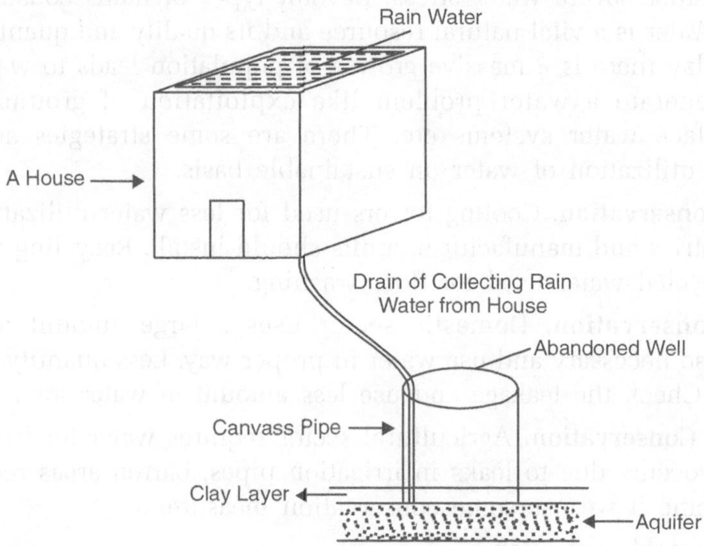

Fig. 6.1. *Rainwater Harvesting.*

Rainwater quality is affected due to contamination, when rainfall occur in areas where heavy industry or corps dusting is prevalent lead to impurity as rain falling in other areas. Water collected from the roofs of buildings and easily transfer in cities and towns. The flow of rainwater tend from roof gutters to a rainwater storage tank. In this way, that collected water used for various purposes. Evaporation loss occurs when water is collected in open areas. Artificial reaching of groundwater helps in reducing the infiltration of saline coastal water to the aquafresh. Flood water during monsoon into the ground to recharge the sub-surface aquafresh. There are two common techniques used for rainwater harvesting one is storage and other groundwater is a new concept of rainwater harvesting and storage is traditional technique. In the artificial recharging we use several methods like : (1) Recharge with shafts, wells, pits etc. (2) Through water spreading, (3) Roof top collection of rainwater.

Roof top rainwater harvesting is necessary for extinction of water, where the top of the roofs is diverted to some surface tank with a delivery system and used for several purposes. It can also help in maintaining underground aquafresh level.

WATERSHED MANAGEMENT

The management of run off and fall rainfall is known as watershed management. Water stored for aquifer recharge, agriculture and wildlife habitat. Streams can hold back water before it becomes a great flood by the construction of a series of small dams, Wildlife habitat and stock watering facilities available by the formation of ponds by these dams. Conservation of wetlands preserves natural water storage capacity and aquafresh recharge zones. The great critical watersheds in the world are the Himalayas and from this a large hydroelectric power potential can harnessed, when proper control measures are taken. For the watersheds protection minimizing ploughing and forest cutting on pump slopes. In this way a watershed in all the land area that drains a particulars streams or river-Bernard, example–Nepal.

RESETTLEMENT AND REHABILITATION OF PEOPLE

The standard and quality of living of the people of a country arise by the economic development. These are different types of natural and man made disasters force people to move out of their land. The most serious effect of socio-economic due to large scale displacement of local inhabitant from their homeland all this due to bigdams. There are more than 20 million people are estimated to have been directly or indirectly affected by big dams. Different types of economic and environmental conditions causes several problems, when rehabilitation occur at out world places, problems such as—loss of kinship system, loss community life, completely destruction of society networks, total loss of productive units and source of income. The resettlement and rehabilitation policy is to ensure that the people displace by dam receivers benefits from it. The objectives of this policy are given below.

1. Employment guarantee to all inhabitants.
2. For the affected population the land, house infrastructure and other type of compensation should be immediately provided.
3. Displacement is necessary, when displaced people should be compensated for their actual losses and supported during the transition period in the resettlement site.
4. Community involvement in planning and implementation of resettlement should be encourages.

The strategies formulated for resettlement and rehabilitation of those displaced by natural and man-made disasters. So the awareness of these negative impacts are necessary, because these occur at frequent intervals. Today adhoc relief measures are taken to cope with the number of problems caused by such disasters.

Case Studies. The main objectives of Resettlement and Rehabilitation policy being implemented by the states in India are as under:

1. Be fully integrated in the community in which they are resettled.
2. Improve the standard of living or at least regain the standard of living the PAFS (Project Affected Families/Outers) has been enjoying prior to their displacement.
3. Ensure adequate participation by the PAFS in their Rand R. The project-wise progress in the area of resettlement and rehabilitation of project affected families of some of the ongoing project in India.

ENVIRONMENT ETHICS

'Ethics' is derived from Greek word *Ethos*, which has a broad mean as character of a person as described by his or her actions. The main theme of ethics like environment degraded by cultural, religious and ethical attitude of man. The various issues, principles and guidelines are associated to the human interactions with their environment.

There are different types of ethics:

1. **The Conservation.** In which its our duty for conserving natural resources and stabilize a balance between resource use and resource availability.
2. **The Development.** It is depend on the action and is inevitable.
3. **The Preservation.** The natural environment gives a beautiful place, so it is our duty to maintain for enjoyment, meditation, breeding etc. and some rare and endangered species must be preserved. But todays several countries facing a number of environmental problems that threaten not only human health and growth of ecosystems. Environmental ethics emphasis on environmental protection acts, but also change in our daily life and the social system must change to become compatible with the global ecology. An exhaustible resource highly requires to regain a balance. Balance by reduces the needs and used material must be replenished.

Need to Change Attitude. Environmentalists suggested the following ethical guidelines to worlds environment:

1. Nature does not exist for human use but also for all species on the Earth.
2. The main aim of man that protect all species and to share earth's limited resources.
3. The wasteful use of natural resources leads to its depletion and determined to the quality of life of other species.

Only better awareness can be created various aspects of environment degradation, as (1) pollution and environment decay, (2) preservation and protection of environment, (3) Sustainable earth.

With the help of these attempts we can change a person into an emotional as well as an intellectual environmentalist. An individual conscious of his environment should try with the following as the individual level by changing his own habits.

1. Dead bodies should not be thrown either into open space or river.
2. Animals bathing in same water prohibited.
3. Proper management of domestic waste is to be practiced in order to avoid pollution.
4. Recycling of domestic waste material.
5. Vegetation covers around your house.
6. For managing domestic wastewater, drainage system properly maintained.
7. Stop the smoking and drinking all other types of intoxicating materials. For the long terms solution of this problem is that we will come from the proper use of science and technology advancement without harming the environmental ethics. The awareness of the need for survival has developed into what we now call the environmental ethic.

WASTELAND RECLAMATION

A wasteland occur due to various constraints and that is at present lying unused or land which is not being used, for example barren land, hilly slopes, overgrazed pastures, degraded forests, waterlogged marshy lands etc. The main causes of formation of wastelands are overgrazing, some side effects of development projects, unscientific land management and misuse. Wastelands may be classified into two types:

1. **Culturable.** Culturable wastelands means that are cultivable and used for agricultural and pastoral purposes and can be afforested. For example saline lands, gulled lands etc.
2. **Unculturable.** These are barren land and not used for cultivation purposes. For example, steep sloping areas or snow covered area, rocky lands, etc. The wastelands has a high significance like disturbance of ecological balance, because wasteland formation and the ecosystem may be directly or indirectly affected by this land. Culturable waste lands can be reclaimed by adopting following remedial measures.

   (1) Proper land use planning, (2) Developing awareness about the potential toxicity of chemicals like pesticides fertilizers etc. (3) Need of development of land protection policies. (4) The hazards related to soil pollution and soil degradation require awareness.

Wastelands covers a large area in each country, there is a great demand for its useful and use by several ways. The main objectives of wasteland are : conservation of biological resources, improve the irrigation facilities, prevent soil erosion, land slides and flooding.

Global Warming

The greenhouse effect is responsible for global warming. At the lower levels of the atmosphere greenhouse gases act like the glass of a greenhouse. That glass are opaque to the heat radiated by the heated earth and trap them. If large amount of CO_2 is present in atmosphere, it cause more heating up of the earth's atmosphere. In this way the temperature of the earth's atmospheres rises. On the global scale, this effect is called as global warming.

Causes. There are various causes of global warming:

1. Main greenhouse gas CO_2 is released by decaying plants as well as human activities, deforestation, combustion of fossil fuels, volcanoes, ocean etc.

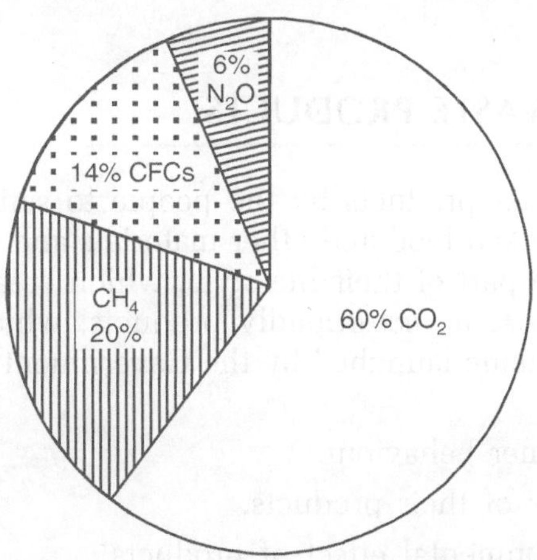

Fig. 6.2. *The Relative Concept Contribution to Different Gases to Global Warming.*

2. Some other gas which are responsible for greenhouse effect like methane, chlorofluorocarbons, water vapour. Where CFC produced by use of coolants in refrigerators, plastic foam materials like 'Thermocoles" propellants in aerosol sprays.
3. Methane is released from coalmines, decomposition of organic matter in swamps, rice cultivation, guts of termites in forest debris.

Effects. Global warming posses the following serious environmental threats:

(*i*) *Sea Level Rise.* When the temperature rises the ocean will warm and expand causing sea levels to rise and flood many of the world's highly productive detail areas. Melting of ice caps and glaciers due to global warming, also contribute to rise in sea level.

It also has an impact on marine ecosystems. The flooding of many of the world's coastal wet lands as the oceans rise would mean the loss of nurseries for many fish, shrimp and birds. Infectious diseases also spread due to sea level rise. Groundwater quality also contaminated with sea water as the sea moves in land and create a drinking water problem.

(*ii*) *Climate Change.* Climate changes will have important effect on agriculture potential may increase or decrease, these could be signify regional declines in productivity, including in Brazil, Central Asia, South East Asia etc. An International Negotiating Committee for climate change was set up on 1990 to formulate a framework convention between countries, to form the basis for schemes of intention regulation of global warming. All the major factors, human and natural are known to affect climate.

There are some other environmental effects as :

1. Several plants species is susceptible to high temperature and leading to rapid loss of biological diversity and resultant loss of natural resource of greater economic potential.
2. Global warming leads to depletion of the ozone layer.
3. Increase in the UV radiation could harm phytoplanktons to an extent where the whole ecosystem may collapses over the globe.
4. One degree of rise in temperature to be equivalent to a change in latitude of 100 km.

CONSUMERISM AND WASTE PRODUCTS

The consumption of consumer products by the people to satisfy their own requirement and lead a more easy life. Packed food and other materials and easily available and people in turn make consumption a part of their life style, which create a soil waste problem in future generation. There are some eco-friendly products, which are easily identified by ecomark an eco-labelling scheme launched by the Government of India. The objectives of this scheme are:

1. Affecting the consumer behaviour.
2. To reduce the effects of their products.
3. Indicating the environmental effect of products.

Consumers are collect the things that they buy. A self-storage industry is in America. This use and throw policy produces an environmental problem. Different types of waste products like cartons, bottles, cans, etc., are thrown each year in the developed countries. Green marketing is an area related to environmental protection. The high consumerism lead to the low incentive to manufacture long lasting quality products and the greater the likelihood that cheaply made products will instead be imported from the lowest-wage, environmentally in regulate overseas manufacture. Consumerism interferes the workings of society by replacing the common sense desire for sufficient supply of life's necessities, community life, a stable family and healthy relationships with an artificial on going and unending quest for things and the money to by them with little regard for the true utility of what is buy. In this way materialism is one of the end results of consumerism.

ENVIRONMENT PROTECTION ACT

The act to provide protection and proper improvement of environment. This act defines the various terms as given below:

1. *'Environment'* consists of water, air and land and the interrelationship, which exists among and between air, land and water and human beings, other living organism, plants, micro-organisms and property.
2. *'Environmental pollution'* – means the present of any environmental pollutant in the environment.
3. *'Environment pollutant'* – means the presence of any environmental pollutant in the environment.
4. *'Hazardous substance'*, means any substance which is harmful to human beings, plants, micro-organisms, our environment and property due to its chemicals or physico-chemicals properties of handling.

The Environment Protection Act, 1986, has different types of objectives:

1. Establish a balance between human beings and their environment.
2. Environment protection and improvement.

The Central Government has powers to protect and improve environment. Procedures, safeguards, restrictions and prohibition for the handling of hazardous substance. Materials and substances, which causes environmental pollution, examined under manufacturing processes. The Central Government, has the following measures:

1. Environmental laboratories and Institutes established and recognized.
2. The standards for the environmental pollutants which discharged from various source.
3. For environmental pollution problems carried out research and investigations.
4. The Central Government under Section 4.3 of this Act may constitute an authority for the purpose of performing such functions and powers.

Section 7.1 suggest that any environmental pollutant in excess of prescribed standards to discharge any operation, industry or process is strictly prohibits Section 8.2, the prescribed procedural safeguards taken when any person handle hazardous substance.

Section 9.3, each person is responsible for environmental pollution mitigation under the concerned authorities.

Under the section 10, the Central government officers have the power for failing or obstructing which will be a punishable offence under this Act.

Section 11, the samples of air, water, soil or substance from the factory are collected by the Central Government and its officers for analysis purpose.

Section 12, the Central Government has the power to establish environmental laboratories. According to Section 16, the company's directors and principal officers fixed the criminal liability in case of an offence being committed by a company.

The various provisions of the Environment (Protection) Act, 1986, have been amended for the effective implementation of this Act.

AIR (PREVENTION AND CONTROL OF POLLUTION) ACT, 1981

According to this Act air pollution means the presence of any air pollutant such as solid, liquid or gaseous in the atmosphere in such concentration as may be injurious to human beings, plants or property. The Air Act, 1981, provide prevention control and abatement of air pollution. For controlling of air pollution boards are establishment. An integrated approach for tackling the problems related to pollution. Under the Water (Prevention and Control of Pollution) Act, 1974, this Act provides the Central Boards for the prevention and control of water pollution. Any fuel which is likely to cause air pollution in any air pollution control area the State government has power to declare air pollution areas and to prohibit the use of it. World's of air pollutants emission form automobiles.

Section 21.3, any person who establish or operate any industrial plant in an air pollution control area, first consult to State Board. For defining the powers, funds, penalties, procedures and functions of Pollution Control Board the Air Act has provisions which ensure the emission standard from automobiles.

Section 24.5, the state boards provide power for testing and examining any control equipment, record, register, industrial plant, documents etc. The State Board entrusted to him, who performing any of the functions.

Section 26, samples of air on emission from any chimney, flue or duct etc., and transfer the same for analysis to the laboratory established or recognized by the State Board. The penalties for violations of the provisions under this Act are the same as under the water Act. The main aim of this Act, before taking him to court as by the time the case is filed in court, the defaulter giving 60 days notice. In which case it becomes difficult to prove the offence, the offender may destroy the evidence. Another drawback is that no permission is reward to be taken from the board for establishing an industry outside the air pollution control area, although its emission may be reaching the air pollution control area too.

The purview of the Act excluded of any air pollutant discharge into the atmosphere by a ship or aircraft's.

WATER (PREVENTION AND CONTROL OF POLLUTION) ACT, 1974

The Water Act provides prevention and control of water pollution. Establish the boards for the prevention and control of water pollution and maintain or restore the wholesomeness of water.

Water pollution defined as the contamination of water or the alteration of the physical, chemicals or biological properties of water or the discharge of any sewage or industrial effluent cause a nuisance or render the water harmful or injurious to human health and safety. Industrial, agricultural, domestic and commercial fields also affected by this. Some other animals or plants life or aquatic organisms have a adverse effect in relation to water pollution. Section 3, the Central Board has the following members:

1. Member-secretary having knowledge and experience of scientific engineering aspects of pollution control.
2. To represent the companies or corporations two persons nominated by the Government.
3. A full time Chairman having practical experience or knowledge about the substance which related to envionmental protection.
4. The Central Government nominated five officials of Central Pollution Control Board for prevention and control of pollution training programs organized. Prepare manuals for disposal and treatment of sewage and effluents.

The State Pollution Control Boards are governed by the directions of Central Pollution Control Board and advises the State government with respect to the location of any industry that might pollute a stream or a well, Central or State Government have power for removing a member of the central or state board at any time by giving him reasonable notice. A member addressing his resignation to the chairman and the chairman may resign by addressing his resignation to the government. One meeting must be held in every three months according to Section 8. Central board performing the following functions:

1. For analysis of water samples to establish or recognise laboatories.
2. To organise training of persons and comprehensive programme for population control.
3. In different areas of the state to promote cleanliness of wells and streams.
4. For testament and disposal of sewage and trade effluents to prepare manuals or guides.

When domestic waste discharges into the water that lead to contamination and causes some water borne diseases like—cholera, dysentery, typhoid etc. The enforcement of suitable preventing measures should have the highest priority for prevention of pollution for drinking-water source so prohibition of discharges both treated and untreated, into water bodies used for drinking and bathing purposes. If a person causes pollution by disposal of polluting matter into a stream or on lands, the Board may approach the court for restraining.

WILD PROTECTION ACT—THE WILDLIFE (PROTECTION) ACT, 1972

It provides protection and conservation of wildlife and to preserve and maintain the biodiversity, ecological processes and life supporting systems. The two new types of reserves—conservation reserve and community reserve are setup in the Wildlife Act.

'Wildlife' means any type of animal, birds, plants, aquatic and land vegetation which forms part of any habitat.

'Hunting' means to capture, traps, kills and poisons any wild animal.

'Habitat' where any type of living being or vegetation live like land, water or vegetation which is the natural home of any wild animal. The Central Government may appoint a Director of Wildlife Preservation according to Section 3, but the State Government may appoint a Chief Wildlife Warden or Wildlife Wardens under the Section 4. The State Government adjacent to national parks and sanctuaries for protecting the habitat of flora and fauna and landscape, seascape etc. The main purpose of conservation reserve and community reserve are improving the socio-economic conditions of the people living in areas as well as conservation of Wildlife. Under this Act several types of conservation projects for individual endangered species like Tiger, Lion, Crocodile were stared. A Wildlife Advisory Board, consist of different types of members as–The Forest Officer Incharge of the State Forest Department, Chief Wildlife Warden, two members of the State legislature etc. Hunting is strictly prohibited by any person under the Section 9, but section 11 and 12 provided hunting in certain cases as:

1. When any wild animal has become dangerous to human life or diseased as to be beyond recover and the Chief Wildlife Warden is satisfied and give order in writing and stating the reason, permit any person to hunt such animal.
2. When any wild animal killed or wounded in good faith of any wild animal in self-defence shall not be an offence.

Section 12, where the Chief Wildlife Warden may grant a permission to any person for hunting any wild animal in certain specified purpose like scientific research and management, education and for collection of specimens etc.

Any licence granted under the Section 44, may suspend or cancel by the Chief Wildlife Warden or Authorized Officer after collecting the reasons in writing and alter giving the licence holder reasonable opportunity to be heard. A person shall be punished with imprisonment for three year or a fine of Rs. 23000/- or sometime both, if he violating any provision of this Act shall be guilty of and offence against this Act.

FOREST CONSERVATION ACT, 1980

Forest is a vital nature resource and gives wide-spread products, which are useful for our life. So its protection and conservation is a necessary fact, which ensure judicious use of forest products.

"Forest" is a community composed predominantly of shrubs, trees and woody climbers.

"Forest-Offence" means an offence punishable according to this Act.

"Forest-Officer" The State Government appointed a person who carrying out the duties and objectives.

Forest Conservation Act enacted by parliament in the thirty first year of the Republic of India come into force of the 25th October, 1980. The State Government may declare that any tree in a protected forest is reserved and the breaking up and clearing for cultivation, for buildings or bearding cattle strictly prohibited under the Section 30. Any person who commit any type of offence, the penalty in term of six months or more, five hundred rupees or more or both as a fine from him will be given under this Act. The offences are given below:

1. Any reserved tree damaged by cattle.
2. Six months or with fine of five thousand rupee or both may extend if a person who kills or catches elephants in any one.
3. Clearing for cultivation, quarrying any stone, burning any lime or charcoal or collecting or removing any forest products are strictly prohibited.

Made rules or carrying out the provisions of this Act by notification in the official gazette. The state government made rules for regulating the management and duties for protection and improvement of village forests.

An authority requires following necessities:

1. Keeping in view the extent of environmental pollution and its hazard to study, plan and implement long-term requirement of environmental safety.
2. Review of legislation.
3. Co-ordinate the activities of other agencies.

ISSUES INVOLVED IN ENFORCEMENT OF ENVIRONMENTAL LEGISLATIONS

Environmental laws in the form of Acts are necessary for protecting and improving our environmental quality. Several types of Boards were setup in various sectors like the Central Government set up a Central Board for the Prevention and Control of Water Pollution, for the Air Prevention and Control the Central and State Pollution for Control Board etc. These Acts was adopted by the state legislatures. The Environment (Protection) Act, 1986 which is called as umbrella of legislation. The "Environment Protection and Improvement" is a responsibility of Central Government but there are some drawbacks in environmental legislations and problems in their effective implementation. The Act empowered the authorities for specific fields which include the following:

1. Taj Trapezium Zone Pollution Authority.
2. Coastal Zone Authority.
3. Environment Pollution Authority, 1998.
4. Loss of Ecology Authority, Tamil Nadu.

There are some pollution Acts has demerits these are given below:

1. The damage caused by the big industries due to pollution is vey significant as compared to the provision of penalties in the Act.
2. The sound policy pronouncements are not backed the pollution control laws.

The pollution control authorities are empowered to impose fines depending on the nature and extent of pollution caused. The Pollution Control Boards are set up for this purpose. The Pollution Control Board have the responsibilities for enforcement of various environmental regulations in addition to the Acts relating to water and air pollution control. This play a vital role with the help of investigation, suggestions, for remedial measures and monitoring their implementation. For effective functioning of the pollution control machinery requires the professional man power and laboratory infrastructure for pollution monitoring.

PUBLIC AWARENESS

Money other most pressing environmental issues concerned not with the natural resources themselves, but with their management, which is largely an institutional and social matter.

People must be made aware about both formal and informal education. Formal education is for the student in the educational system, non-formal education is for the non-educational institutions. The University Grants Commission recently introduced syllabus on environmental studies for the undergraduate courses in all streams.

Public awareness can be created by following method:

1. **Among the People through Mass Media.** Awareness created by radio, newspaper, magazines, telling environmental conservation stores audio and video CDs etc.
2. **Among Student through Education.** Student awareness created by teaching environment as compulsory subject from primary classes to graduate classes.
3. **Among the Formers and Local People.** The seeds of environmental awareness can be effectively propagated by organizing training camps.
4. **Organization of Environment Movement.** In critical situation when environment degradation takes place, environmental movement are organized such as Chipko Movement and Silent Valley Movement and so on.
5. **National Policy on Environment.** The worldwide funds for nature WWF, center for science and environment, CSE and Bombay Natural History Society, BNHS, are all government organization and play a significant role in environmental awareness through research and other activities.
6. **Among the Politician and Beuracrates.** As is play an important role in bringing movers in the society. These people should attend orientation courses or workshops related to environmental awareness.
7. **Non-Government Organization.** Non-government organization play a significant role in creating a like between government and local people for addressing environmental issues and generating an awareness. There are about 200 Non government organization out of which 130 are engaged in awareness programmes, 50 in pollution control, so in nature conservation 45 in afforestation and social forestry, 15 in rural development and 10 in eco-development. All these are academically-oriented activities.

ACID RAIN

When fossils fuels burn that produces oxides of N_2 and S_2. Then these two elements combine with atmospheric O_2 and form NO_2 or SO_2. After this, they combine with rain water and form H_2SO_2 or HNO_2. This acid rain is harmful to environment directly. Acid rain affecting the flora and fauna and increase soil acidity. Sometimes they create a acidification problem in lakes and streams and disrupt the aquatic life, human health, buildings, bridges, monuments also affected by this. Acid rain is mainly responsible for yellowing of Taj Mahal.

Causes

Combustion of fossils fuels emit a large number of pollutants as nitrogen and sulphur, internal combustion of automobiles produces CO_2, hydrocarbons and CO. Industrial sector emit a number of pollutants like CO_2, SO_2, NO_2, in the environment. Some other sites also produces oxides of sulphur and nitrogen as power plants, domestic fires, smelters etc. These different types of pollutants combine with water or absorbed by moisture forms of acids and known as acid rain. In this way acidification of rain water is a man-made phenomenon.

Effects

Acid rain affected the land as well as aquatic life in lakes, streams. Its adverse effect on vegetation, forests, crops and reduces the agricultural productivity. The main harmful impact on monuments like yellowing of white marble due to acid precipitation.

$$CaCO_3 + H_2SO_4 \longrightarrow CaSO_4 + CO_2 + H_2O$$

(Acid Rain)

$$CaCO_3 + 2HNO_3 \longrightarrow Ca(NO_3)_2 + CO_2 + H_2O$$

Acid rain causes damage to rocks and increase rate of weathering of rocks. It has a corrosive effect on metals in the form of rusting iron. Vegetation damage occur in the form of suppression of growth, leaf drop and reduction the yield. Increased leaching of nutrients as calcium due to acidity of the soil, Acid or Base catalysts are responsible for changing of metabolic rates of organisms. Acid pollution is damage the human nervous system. Industrial area is largely affected by acid rain where the pH value below to the critical value recorded in Delhi, Bombay and Calcutta.

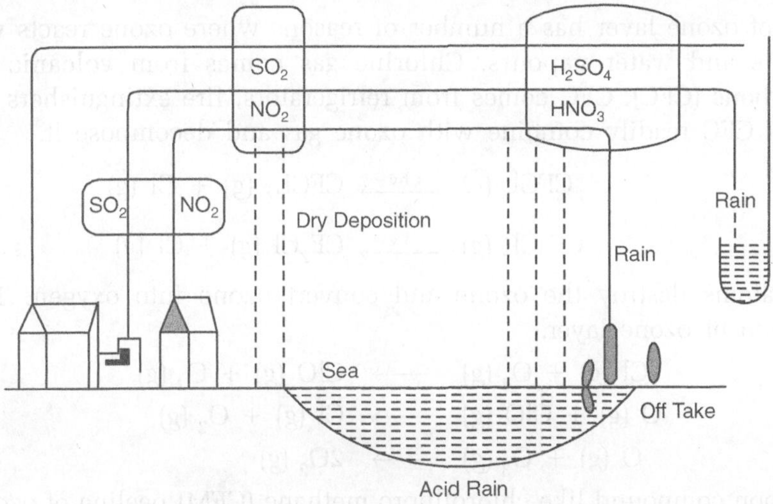

Fig. 6.3

OZONE LAYER AND OZONE HOLE

Depletion of the Ozone Layer

Ozone layer occurs in the stratosphere region. Ozone layer filter the UV radiation and thus protects us from its harmful effects. In the stratosphere the temperature increases at a high rate. But its depletion is a major cause of polar regions and cold climate is one effect of this. The harmful radiations produces sun burns, skin cancer, leukamia, edema, skin swelling, lung cancer and sometimes lead to death.

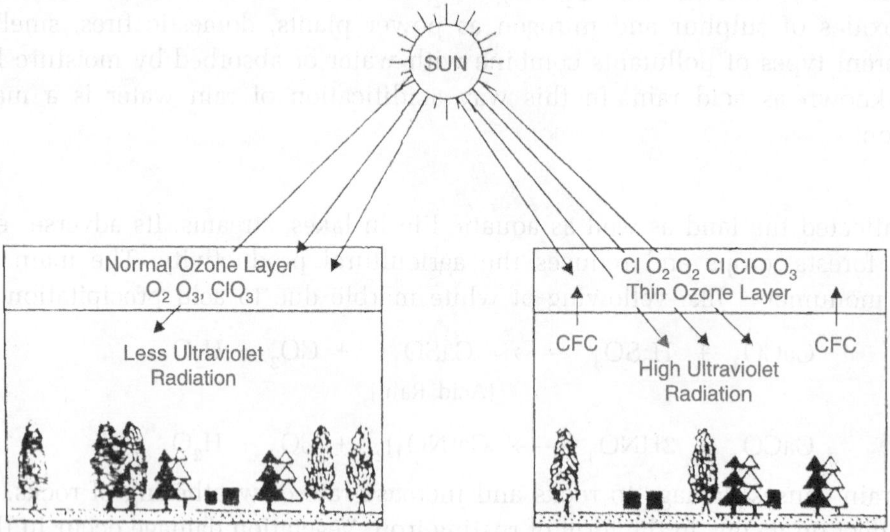

Fig. 6.4. *The effect of man-made chemicals on the ozone layer in the stratosphere ; Left : less UV-radiation reaches the earth's surface due to intact ozone layer ; Right : high amount of UV-radiation reaches the earth's surface (ozone hole) due to thinning of ozone layer by the ozone-depleting substances.*

Causes

The depletion of ozone layer has a number of reasons where ozone reacts with nitrogen oxides, chlorine and water vapours. Chlorine gas comes from volcanic activity and chlorofluorocarbons (CFC). CFC comes from refrigerators, fire extinguishers, arosol spray propellants etc. CFC readily combine with ozone gas and decompose it.

$$CFCl_3\,(g) \xrightarrow{hv} CFCL_2\,(g) + Cl\,(g)$$

$$CF_2Cl_2\,(g) \xrightarrow{hv} CF_2Cl\,(g) + Cl\,(g)$$

Chlorine atoms destroy the ozone and convert ozone into oxygen. This leads to gradual depletion of ozone layer.

$$Cl\,(g) + O_3\,(g) \longrightarrow ClO\,(g) + O_2\,(g)$$
$$O\,(g) + ClO\,(g) \longrightarrow Cl\,(g) + O_2\,(g)$$
$$O\,(g) + O_3\,(g) \longrightarrow 2O_2\,(g)$$

Fluorocarbon compound like chlorofluoro methane (CFM) peeling of ozone umbrella.

Importance

Ozone layer is useful for us and check the UV radiations coming from the sun. The umbrella of ozone plays a key role in the climatology of the earth. It can help in controlling the thermal budget of the earth. The amount of ozone is related to life-sustaining process. Its depletion has a sever impact on vegetation, health and life of human being etc.

Effects

Ozone layer depletion has a wide-spread effects:

1. Temperature increase around the earth causes warming of the earth leads to melting of polar ice-caps and flooding of coastal regions.
2. Ozone is extremely toxic and increases susceptibility to infection.
3. Ozone is alone harmful to vegetation and crops. White pins needles are also affected with tipburn caused by ozone.
4. The main injurious effect of ozone due to lipid oxidation. Increase in UV radiation reaching on the earth's surface cause a 30% increase in non-fatal skin cancer.

OZONE HOLES OVER ANTARCTICA

Scientist discovered a hole in South Pole. The hole in ozone layer as large as that of the United State and this was followed by Montreal Protocol in 1987. CFC and other chemicals depleting the ozone. US satellite taken photographs. These shows a big hole in ozone layer over the Antarctica, which increase about 3% CFC affecting at an alarming rate so USA immediately banned the use of CFC. Montreal Protocol has a main purpose in reduction in the global production of the CFC.

EXERCISES

1. Explain recent approaches to sustainable development.
2. Discuss the urban problems related to energy.
3. Write a short note on water conservation.
4. Explain rainwater harvesting.
5. Discuss the problems and concerns of resettlement and rehabilitation of people ?
6. What are issues and possible situations related to environment ethics ?
7. Write short notes on these:
   (*a*) Climate change (*b*) Global warming
   (*c*) Ozone layer (*d*) Acid rain.
8. Write short notes on these:
   (*a*) The Air Act, 1981 (*b*) Water Act, 1974
   (*c*) Wildlife Act, 1972 (*d*) Forest Act, 1980.
9. Write a short note on ecological forming system ?

Unit **VII**

HUMAN POPULATION AND THE ENVIRONMENT

POPULATION GROWTH

Human population was approximately 0.6 billion in 1700 A.D. At the beginning of the twentieth century, it increased to 1.6 billion and by the end of the century, the human population stood at 6.1 billion. This sudden increase in population size over a relatively short period is known as **population explosion.** Currently, the pollution is increasing by more than 80 millions per year, *i.e.*, every 12 or 13 years, a billion people will be added. This rapid growth of population is the result of the spread of public health programmes throughout the developing countries and the rise in good production after second world war as a result of the 'Green revolution lain infant mortality' Population even will grow more in twenty first centuries. The next century's growth will occur almost exclusively in the developing countries. India's population has reached 1 billion on 11th May 2000 and the country is likely to become the world's most populous country by 2050. The population estimates for major areas and regions of world from 1980–2050, and the projected population for 2050 of first fifteen countries are shown in the below mention tables:

Table 7.1. Population Estimated from 1980–2050 for Major Area/Regions of the World

(Fig. in Million)

Area/Region	*1980*	*1985*	*1990*	*1995*	*2000*	*2025*	*2050*
World	4453	4842	5248	5677	6067	7810	9039
Africa	476	553	645	753	800	1258	1804
N.America	252	263	275	287	306	374	444
Latin America	362	406	453	501	482	657	772
East Asia	1182	1252	1317	1390	1493	1669	1585
South Asia	1408	1752	1740	1909	2003	2754	3287
Europe	484	492	499	505	827	714	658
Oceania	23	25	27	29	31	39	44

Source. *World Population data sheet, 2000 and based on medium estimates and projections of United Nations assessed in 1982.*

Table 7.2. First Fifteen Largest Population Countries (Projection for 2050)

Rank *(in million)*	*Country*	*Population*
1.	India (Asia)	1628
2.	China (Asia)	1369
3.	United States (N. America)	404
4.	Indonesia (Asia)	312
5.	Nigeria (Africa)	304
6.	Pakistan (Asia)	285
7.	Brazil (South America)	244
8.	Bangladesh (Asia)	211
9.	Ethiopia (North-East Africa)	188
10.	Congo, Democratic Republic of (ZAIRE) Middle Africa)	182
11.	Mexico (North America)	152
12.	Philippines (South-east Asia)	140
13.	Russia (Europe-Asia)	128
14.	Vietnam (South-East Asia)	124
15.	Egypt (Northern-Africa)	117

Source *: World Population data sheet, 2000*

VARIATION AMONG NATION

The distribution of people on the earth is not equal more than half of the world's people live in Asia. Therefore are some areas, which is densely populated as China, India, Western Europe and north eastern USA. Good land area, excess of natural resources, extension of railways, roads and other modes of transportation are necessary for settlement. Some other factors which disturb the settlement like draught, cold, poor quality of soil, climate, heat etc. Only one twenty fifth of the total land area occur in Europe where about one-ninth of the world's people live. Quarter of the land surface occur in African continent where over one eighth of the world population live. Northern European countries have less number of people than other European countries.

POPULATION EXPLOSION

The rapid increase in world population is known as population explosion. The developing countries has a higher increase rate than the development countries. So there adopt population control policies. The increase in population growth produces major problems in very poor, agricultural countries. Population control is most important priority in the densely populated countries, but childhood deaths and short life spans used to limit population growth.

When there is high birth rate and a low death rate is creating a rapid population increase in many countries in Asia. Death rates have fallen faster than birth rates. Although productivity increase in general more quickly than population, growth at the world level but should not lead to the exhaustion of resources. There are several causes of rapid population growth:

1. There were great advances in science and technology, the success in reducing death rates by several factors—(*a*) Betterment in the human health. (*b*) Development of medical technology. (*c*) Increases in food production and distribution.

The children's are susceptible to common disease like measles or the flu. But today advance technology and different kinds of food which could produces by people, these leads to low death rates. People were able to fight and cure deadly germs that harmful and killed them. Low death rate and improved life quality due to these discoveries and inventions. The poor countries have a high populations growth. Increased death rate and disease are closely related to overpopulation and poverty. The end of world war-II has the expansion of population size and this is a main cause of their poverty. Poverty is a condition of chronic deprivation and need at the family level. The lack of basic requirements produces physical weakness and poor health. In this way poor health makes the human for poor working and put them into deeper poverty. There is high struggle occur for food, water, shelter, educational and population explosion occure due to health case and improved sanitation leads to low death rate than the birth rate. Our country has led to serious overcrowding problem where population grows about twice as fast as the population of other countries. The rapid population growth is also related to a massive rural-urban migration. This migration from countryside to city is producing a serious social problem as increase in crime rate, unemployment etc. The earth's life support systems have been depleted due to over population growth rate. The school level education develop more awareness about population issues in order for them to make projections about the future generations. The strategies and action plans are helpful for a successful solution to many populations problem. Health care, women's education and family planning build a basis for action. So there is only one way to raise economic standards by the reduction of birth rate.

FAMILY WELFARE PROGRAMME

It has been accepted worldwide that family planning enhance individual's health and welfare. The Government of India was the first one to recognise the importance of family planning and initiated Family Planning Program in 1951. Since the introduction of the programme in 1951 till now, a substantial improvement has been achieved in the planning of families including the health of women and children. The total fertility rate (TFR) has declined from 6 to 3.3, the infant mortality rate from '146 to 72 per 10000 and crude death rate 25.1 to 9. However, reduction in the birth rate has been less steep, declining from 40.8 m 1951 to 22 m 2000.

The country's population policy is more popular than realistic and effective. Women are being targeted for population programmes, while men evaded the whole issue. This is one of the examples of discrimination in applying the principles of population policy. Recently, an integrated approach rather than target-oriented programmes, concentrating on child mortality, maternal safety and the endowment and education of women was introduce, but it has also failed to make a dent because the policy remained largely on papers.

National Population Policy, 2000

The National Population Policy, 2000 (NPP, 2000) affirms the commitment of the government towards voluntary and informed choice and consent of citizens while availing of reproductive health-care services, and continuation of the target-free approach in administering family planning services.

The immediate objective of the NPP 2000 is to address the unmet needs for contraception, health care infrastructure, and health personnel, and to provide integrated service delivery for basic reproductive and child health care. The medium-term's objective is to bring the TFR to replacement levels by 2010, through vigorous implementation of inter-sectorial operational strategies. The long-term objective is to achieve a stable population by 2045, at a level consistent with the requirements of sustainable economic growth, social development and environmental protection. In presence of these objectives, the following National Socio-Demographic Goals to be achieved in each case by 2010 are formulated:

1. Address the unmet needs for basic reproductive and child health services, supplies and infrastructure.
2. Reduce maternal mortality ratio to below 100 per 100,000 live birth.
3. Reduce infant mortality rate to below 30 per 1000 live birth.
4. Promote delayed marriage for girls, not earlier than age 18 and preferably after 20 years of age.
5. Make school education up to age 14 free and compulsory, and reduce dropouts at primary and secondary school level to below 20% for both boys and girls.
6. Achieve universal immunization of children against all vaccine preventable disease.
7. Achieve 80% institutional deliveries and 100% deliveries by trained 20 years of age.
8. Achieve 100% registration of births, deaths, marriage and pregnancy.
9. Achieve universal access to information/counselling and service for fertility regulation and contraception with a wide basket of choice.
10. Prevent and control of communicable disease.
11. Bring about convergence in implementation of related social sector programmes so that family welfare becomes a people centred programme.
12. Promote vigorously the small family norm to achieve replacement levels to the Total Fertility Rate (TFR).
13. Integrate Indian Systems of Medicine (ISM) in the provision of reproductive and child health services, and in reaching out to households.
14. Constrain the spread of Acquired Immuno Deficiency Syndrome (AIDS), and promote greater integration between the management of Reproductive Tract Infections (RTI) and Sexually Transmitted Infections (STI) and the National AIDS Control Organization.

If the National Population Policy (NPP) 2000 is fully implemented, we anticipate a population of 1107 million in 2010, instead of 1162 million projected by the Technical Group on Population Projections.

Environment and Human Health

Family Welfare programmes is to support well being of the people. It was launched in 1951 in India. The main aim of this programme is reducing the birth rate for stabilization of population level. It has the benefit of the small family norms and removal of socio-cultural barriers for adoption of family limitation programmes. This programme is known as a priority area and a 100% centrally-sponsored programme. This programme has list of family planning methods. The common activities of this programme are family education, Health facilities, planning etc. A network of sub-centers of Community Health Centers and Primary Health Centers in the rural areas, hospitals and dispensaries in the urban area through Family Welfare services are provided to the community. The Minimum Needs Programme set up this network.

1. **Education.** An educated person understands the need of welfare and can differentiate between good and bad. Thus education is a basic requirement of human beings.
2. **Family Planning.** Family planning programmes are different from society to society where they includes information about abortion and sterilization, sex education, methods of birth control etc. Educational and medical services are necessary for couples in making small family. Family Planning Programmes or methods are helpful for birth rates in most of the human cultures have dropped.

There are some methods for controlling human population:

1. Condoms are used by males for controlling pregnancy and protect against disease like AIDS, herpes etc.
2. Sepermicides, diaphragms, abortion in some countries are less popular methods are widely used.
3. Preventing pregnancy by sterilizing females or males if they do not want any more children.

HUMAN RIGHTS

Human rights are the rights which are possessed by all human beings irrespective of their race, caste, nationality, sex and language, simply because they are human beings. Human rights are sometimes called *fundamental rights* or *basic rights* or *natural rights*. As fundamental or basic rights, they are those rights, which must not to be taken away by any legislature or any act of government and which are often set out in a constitution. As natural rights they are seen as belonging to men and women by their varying nature.

Human rights occupy a significant place in the U.N. Charter. The Preamble of the Charter reaffirms faith in fundamental human rights and the dignity and worth of human persona and in equal rights of men and women. It is one of the purposes of the U.N. to achieve international co-operation in solving international problems of economic, social cultural or humanitarian character and in promoting and encouraging respect for human rights and fundamental freedoms for all without distinction as race, sex, language or religion. Besides this, it is provided in the character that the General Assembly shall initiate studies and make recommendations for the purpose of promoting international cooperation in the economic, social, cultural, educational and health fields; and assist in the realization of the human rights and fundamental freedoms for all without distinction as to race, sex, language or religion. Further with a view to the creation of the conditions of stability and well being which are necessary for peaceful and friendly relations among nations based on respect for the principle of equal rights and self-determination of people, the U.N. shall promote universal respect for and observance of human rights and fundamental freedom for all.

The commission on Human Rights was established by the economic and social council in 1946. It acts as permanent machinery for the supervision of the problem of protection to human rights. Under its terms of reference, the commission prepares recommendations and reports on (*i*) an international Bill of Human Rights, (*ii*) International Conventions and declarations of civil liberties : the status of women, freedom of information and other similar matters, (*iii*) the protection minorities, (*iv*) the prevention discrimination on the basis of race, sex, language or religion, and (*v*) other matters concerning human rights.

Human Rights—which is essential for the human being and each will receive respect and protection. A person should have the right to live in a healthy and clean environment. The 'Universal Declaration of Human Rights' adopted by the United Nations on Dec. 10, 1948. It shows a world milestone in the long struggle for human rights. The declaration state that all the human beings are born free and have their own rights.

The man's natural and man made environment are necessary for all well being and to the enjoyment of basic human rights this declaration said in 1972, Stockholm. The relationship between the quality of human environment and the enjoyment of basic human rights are recognized in the United Nations General Assembly, 1986. Environment protection and sustainable development emphasized in Rio Declaration, 1992. Agenda 21 known for the improved living standards for all, the fulfillment of basic needs, better protected and managed ecosystems and a safer, much prosperous future life. Human rights have a wide-spread disparity between the developing and developed countries. The present and future generation have a endanger life due to environmental deterioration. There are some fundamental rights, these are given below:

1. The right to education, conscience, freedom of thought and religion.
2. The right to freedom of expression and opinion.
3. The right to liberty, life; own property and security of person.

According to Article 48A, a Directive Principle of State Policy provides that the State shall endeavour to protect and improve the environment and protect the forests and wildlife of the country. Another expansion of the right to life is the right to livelihood (article-41), which is a directive principle of state policy. There is a close connection between article 41 and 21 was established in the 1980's.

VALUE EDUCATION

Values are ideals, beliefs or norms, which a society or the large majority as a society members hold. They are the part and parcel of the philosophy of a nation and that of its educational system. Education in one of the most important tool in bringing about socio-economical and cultural progress of a country.

Education is based on intellectual, aesthetic, social and cultural, emotional, moral, spiritual and religious values. The value education helps young people to lead more personally satisfying and socially constructive lives. Values education encourages the development of moral codes like honesty, truthfulness and justice.

Value education has different phases *i.e.,* value awareness, value orientation, value appraisal, value selection, value commitment and value action. The basic aim is to create and develop awareness about the values, their significance and role.

Value education is to prevent erosion of value in public life has been a matter of concern, science independence. The value embedded in the Indian Philosophy, cultural heritage and literature need to be nurtured in school. A member of Education commission and committee set up by Government of India have made recommendations about the need for formulation of programmes on education in human values in schools. The S.B. Chavan Committee Report (1999) submitted to the Indian Parliament emphasized the need to nurture case universal value like truth, peace, love, righteous, conduct, and non-violence in student. These universal values have been emphasized by all religion.

NCERT has been identified by the MHRD (Department of Education), Government of India as the nodal center for strengthening value education in the country at school level. A National Resource Center for Value Education (NRCVE) has been set up in the Department of Educational Psychology and foundation of education, the NRCVE was inaugurated on 14th September, 2000. The focus of the programme is on generating awareness, material development, teachers training, and promotion of research and innovation in the area of education of human values and development of guidelines for value education in the school system.

Value-based Environment

The principles of ecology and fundamentals of environment can really help create a sense of earth citizenship and a sense of duty to care for the earth and its resources and to manage them in a sustainable way so that our children and grand children too inherit a safe and clean planet to live on.

How environmental education be made value based:

(*i*) Social values like love, compassion, tolerance and justice, which are the basic teachings of most of our religious need to be woven into environmental education.

(*ii*) Cultural and religious values inherited in Vedas like *"Dehi me Dadamite" i.e.,* "you give me and I give you" emphasize that man should not exploit nature without nurturing her.

(*iii*) Environmental education should encompass the ethical value of earth centric rather than human centric world view. Instead of considering human being as supreme we have to the welfare of the earth.

Value-based environmental education can bring in a total transforming of our mind at our attributes and our life-styles. The value elements in environmental education a love can succeed in achieving the real goals of environmental literacy.

HIV/AIDS

AIDS, the acquired immuno-deficiency syndrome (sometimes called 'slim disease') is a fatal illness caused by retrovirus known as the human immuno-deficiency virus (HIV) which breaks down the body's immune system, leaving the victim vulnerable to a host of life threatening opportunistic infections, neurological disorders of unusual malignancies (WHO, 1986). Strictly speaking, the terms AIDS refers only to the last stage of HIV infection. It can be called as modern pandemic. Affecting both industrialized and developing countries, all over the world.

In May 1986, the International Committee on the Taxonomy named the AIDS causing virus as human immuno-deficiency virus. It is $\frac{1}{1,00,000}$th mm in diameter. It is a protein capsule containing two short strands of genetic material (RNA) and a few enzymes. The virus uses human cells to perpetuate itself.

The virus replicates inactively dividing T4 lymphocytes and like other retroviruses can remain in a latent state in lymphoid cells that can be activated. The virus has the unique ability to destroy human T4 helper cells, a subset of the human T4 lymphocytes. The virus can spread throughout the body. It can pass through the blood-brain barrier and can destroy some brain cells. HIV mutates rapidly. There are two types of HIV—the most common HIV-1 and a more recently recognized HIV-2.

WOMEN AND CHILD WELFARE

Women and children are important components of human resource development. For improving the socio-economic status the department of women and child development has been set up. Lack of proper malnutrition, infection and education these problems affecting the women and child welfare. The department of women and child development of women's potential play a vital role in the development process as beneficiaries. There are different types of policies, which have purpose for women that making them economically independent and self-reliant.

1. **Education.** Many developing countries have the problem of female education. Where they decide that education need only for male child as compared to the female child. So the female education free at the primary level by the several under-developed nations. Our country also has a low rate of female literacy.
2. **Malnutrition.** Today there are several developing countries who facing the problem of malnutrition. Pregnant woman, and children required high nutritious food. Malnutrition has some adverse effects like low birth weight, toxemias of pregnancy, maternal depletion, anaemia etc. There are some nutrition measures which directly or indirectly recover the malnutrition value-distribution of iron and folic acid tablets feeding programmes, improvement of environmental sanitation, family planning, education and food hygiene etc.

There are some programmes set up for the development of women and child:

1. **Maher Yojana.** Under this programme government give shelter for women who come to the government institutions. Several NGO's will be extended to shelter homes.

2. **Annapoorna Yojana.** Annapoorna yojana launched by the government where women to group together in preparing nutritious food for anganwadis are paid for their work at 15% per beneficiary per day.
3. **Mahila Mandal.** Mahila Mandal scheme is important in the rural areas impart; training to the rural women for self-employment.
4. **Bal Sadan.** A group of 18 to 21 Bal sadan is called as Balgram and Bal sadan where one institute has 8 to 9 children is known as Bal sadan and supervised by a house mother.
5. **Mahila Bal Kalyan Samiti.** Zilla parishads have a "Mahila and Bal Kalyan Samiti" under the State government. The children and women welfare programmes were superivsed by this committee. The child development scheme, rural development programme, and other schemes being implemented by the women and child welfare department.

The policy of Government National Children Board was set up in December 1974, which provide care, protection, testament, rehabilitation and development of neglected children.

ROLE OF INFORMATION TECHNOLOGY IN ENVIRONMENT AND HUMAN HEALTH

Information Technology, is important for communicational and dissemination of information for the protection environment. Environment information system has been set up in India by the ministry of environment and forests with its head quarters in Delhi. These center works in area as remote sensing, renewable energy, environmental management, pollution control, biodiversity etc.

Today the Internet or worldwide web available for getting information pertaining to any subject and the area of environment also.

Geographical Information System (GIS) and Remote Sensing is also useful for managing natural resources. Remote sensing used at regional, district and nation levels for several development activities. Different type of natural disasters and changes in the environment can be detected with the help of these technologies. The national level standing committees on bioresearches water agriculture, ocean resources and number of application projects have been taken up. Except this the IT is useful in rural areas.

Role of Information Technology in Human Health. It is associated to all aspects of healthcare and at many different points of contact. Satellite remote sensing for disease control will play a vital role in public health in the near future. It is useful for assisting epidemiologists and areas where disease outbreaks are commonly occur. The capability of using remote sensing date to monitor areas in exact time and predict disease outbreaks. In the developing countries IT is helpful to design of public health systems. Rising costs of healthcare, increased longevity and so costly medicines are the main reason patients, national public health systems. **Apollo Hospitals** and **Escorts** in India to be able to offer super specialisations. IT-enabled tools are helpful for DNA and protein testing.

POPULATION EXPLOSION

Population Explosion is the situation of such population growth in which the limits of tolerance of various existing environmental conditions (the carrying capacity of ecosystem) are over. In other words when the population growth of any country or region reaches a level which causes the country resources helpless to maintain or reform the standard of living of the population is called population explosion. **Harvey Libenstein** has called this situation as low level Equilibrium Population Trap. We can simply explain the conditions in this way also. When the income and production of a certain are became incapable of fulfilling the needs of aspired living standard to the growing population the situation is known to be nominated by population explosion.

Effect of Population Growth on Development

The Ex. Planning Minister, Ashok Mehta once told *"Population growth is like a night thief who steals away our success of progress."* Indira Gandhi also expressed her worry in this respect in the worked : *"In such a situation when rapid growth of population is in vogue to proceed further with the planned development is like to construct the house on such a ground, which as usually flown away by the flood"*. These statements were not baseless because Indians could never be prepared to divert themselves from their conventional living and thinking, However when the over population issue came in view of the developmental aims and objects, birth control and family planning programmes were enforced by the government. But these could not give successful results. Hence the growth in population is a major hindrance in development of India. This condition is affected India in the followings ways:

1. **Fall in Per capita Income.** It is quite apparent that in case population increased percapita income of the country must reduce. Though every possible effort has been made but without any significant result. In the last 40 years national income of India was Rs. 1127 percapita which increased in the year 1990-91 about 2226 percapita. It means that the income percapita increase is quite less in view of the considerable increase in production, due to growth of population.
2. **Population and Capital Formation.** In case of population growth savings are badly affected. The major part of income is spent on goods and services. Consequently there is very slow growth rate of capital formation. Ultimately advancement of industrialization suffers and restrained.
3. **Unemployment.** The problem of unemployment is increasing on account of over population in the country. In 1951 the number of employed persons was 40 lakh, which reached to 5 crore in 1994–95. Nearly 7.5 crore people were given employment during the Five Year Plans.
4. **Increasing Pressure on Land.** Population growth is the major cause of high pressure on land 70% of Indian population earns its livelihood through agriculture. The acres of land per capita has decreased from 0.86 acres per head to 0.65 acres per head. The growth rate in agriculture is very slow, inspite of the fact that encouraging efforts have been made by the government in so many ways but the poverty does not allow the small agriculturists to avail the gains of these efforts.

5. **Heavy Pressure on Economy.** On account of population growth every year 25 lakh new houses are required and 126 lakh schools, associate requirement of 37 lakh additional teachers and 44 lakh additional employment it thus an ultimate sequence. Cloth and food requirement is thus consequential to this progress, which causes heavy pressure on national economy.
6. **Problem of Food.** Naturally, food is the hard and fast need of the increasing population. For a long time the country has been facing this problem. Though in comparison to 1951, the food grains production has increased by three times even then India has to input a lot of food grains worth rupees about 10 thousand crore.
7. **Non-Productive Consumers.** Increase in population is one of the most effective factors, responsible for the growth of non-productive consumers. Every year new children have to take birth. Old age people and the unemployed staff of 15 to 19 years has to increase. This class of people are considered to be unproductive part of population.
8. **Social Problems.** Growth of population is also cause of growing so many social problems. For example, increase in the graphs of criminal activities, litigation and family disputes etc.

PUBLIC HEALTH AND POPULATION

Though population growth is a material problem at presents specially for India and China, but the industrial development and technical devices have emerged as the causes of new trouble to human being. Some times even the well educated doctors have remained unsuccessful in providing curable medical aid to the patients. Under this situation the government duty of taking care of public health is enhanced. However, in the historic ages the government were also dedicated to take care of public health. So many wealthy persons of the times passed through ages of history remained fond of opening hospitals where the poverty stricken people were treated free of cost, the same tradition is however toiling after today. In this situation Government can not ignore their duties in this respect, how serious might be the problem of population growth. At present the whole world is facing serious new health challenges. With faces of time, the death toll form Acquired Immuno Deficiency Syndrome (AIDS) may rise to millions in number annually. The available malaria drug even have become ineffective to malarial parasite. Hence malaria deaths increased to nearly two millions a year within a short time. Increase in human population brought non-communicable disease of adults and the elders. It is estimated that if present smoking patterns remain prevailed in a customary fashion, tobacco affected deaths from heart disease and cancers alone are likely to double by the first decade of 21st century a year.

Now it is not a hidden fact that infectious disease are not a matter concerned to some particular country or region. Their claws are spared all over the countries of the world. Emerging and re-emerging diseases are index of large scale environmental change.

Though at present HIV/AIDS has diseased the third world prominently but no other country has been spared by it. Along with malaria, kala azar, and plague, tuberculosis and recent spread of viral hemorrhage fevers affected the world countries.

The resurgence of plague in India 1994 after three decades of absence from the country is supposed to be the probable result of environmental disturbances, disruption and destruction of the world's natural life supports system constitute the fatal a threats to human health.

Disease can be classified under four categories:

1. Water Borne disease
2. Air Borne disease
3. Food Borne disease
4. Vector Borne disease.

Various vectors such as mosquitoes, flies or animals are the transmitters of so many diseases.

AIDS/HIV

AIDS is most dangerously fatal disease of the present age which breaks down the immune system of body. It give such illness which leaves the victim. Vulnerable to a hash of life threatening opportunistic infections, neurological disorders or unusual malignancies. Medically AIDS and HIV are the two stages of the same disease. Previously any person might infected by HIV (Human Immuno Deficiency Virus) which causes the human body to proceed further to the jaws of AIDS (Acquired Immuno Deficiency Syndrome). It is sometimes called skin disease and neither the developing countries nor the under developing countries or the developed ones that is the foremost industrialized ones are not out of its approach. HIV has infected millions of women, men and children throughout the world.

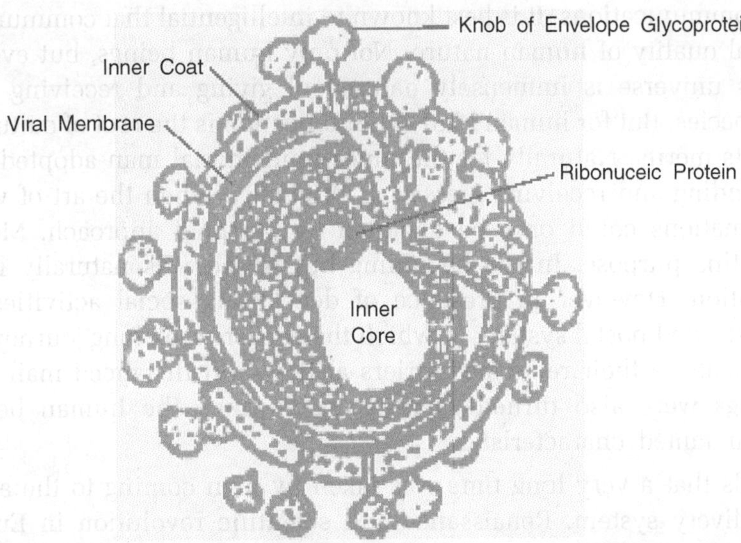

Fig. 7.1. *Three Dimension Structure of H.I.V. Virus.*

AIDS was first recognized in USA in 1981 when it was traced. Retrospective analysis disclosed that is occurred in 1975 in USA and in late 1980s in equatorial Africa (WHO-

1985). An expansion of HIV has recently opened its jaws particularly in Thailand, Burma and India. This fatal disease has also reached Paraguay, Greenland and Pacific Island Nations of Fiji, Panama, New Guinea and Somoas.

In May 1986, the International Committee on the Taxonomy named the AIDS causing virus as human immuno deficiency virus. It is 1/10000th of mm in diameter. The virus uses human cells to perpetuate itself. It has its concentrated abode in blood, semen and CSF. Its concentration have also been found in tears, saliva, breast milk, urine, cervical and vaginal secretion. By this time medically ascertained theory in respect of the AIDS transmission is possible through blood and semen. The real cause of AIDS infection is sexual relationship. The virus of the disease may lay dormant for period of 8 to 10 years. After this it suddenly rises grownup and causes death. But this time no treatment has been found out. The real treatment of this disease lies in taking precaution. A person should not have sexual contact with an AIDS affected person, and he should also not use the same syringes, which might have been used by other persons.

Naturally in the present circumstances when population is tremendously increasing and people of every country are very close to each other the Government are required to take such steps that disease like AIDS etc. might not have an easy way to migrate to their countries.

INFORMATION TECHNOLOGY

We can be amazed in our own deep sense of thinking, if take the view of modern world, because though the present historical age, man did not attain such wonderful devices, as have been achieved by him in the last one century. Most wonderful of them all is the technology of communications. It is best known to intelligential that communicative anxiety is the congenital quality of human nature. Not only human beings, but every life bearing creature of this universe is immensely natured of giving and receiving information in between their species. But for human kind communication is the root of cultural civilization, and it has got its merits. Naturally for the times immemorial man adopted every possible technique of sending and receiving messages. Originally, when the art of writing was not invented, informations could only be departed by personal approach. Messengers were dispatched for this purpose. In the beginning but sound was naturally instrumental to release information. However in presence of developing social activities and political affairs, kings instituted postal system, in which the carriers took long journey on horseback, delivering the mails to their relieving carriers at the short distanced mail posts. Pigeons, hawks, and dogs were also turned as messangers since the human being can to be understand their tamed characteristic.

History tells that a very long time was taken by man coming to the advancement of rapid postal delivery system. Renaissance and scientific revolution in Europe caused a change and as soon as fast means of transportation, naturally could be procured quick measures of communication were adopted. But the scientific progress did not stop, turn by turn it gave wireless and telephone to public, then radio, cinema and television. Now the masses could be sent of the declarations of government through these devices, while the same were to be notified before the gathering of masses caused to be called by beating of drums. Media also came in existence by this time, which took the task of informing public of not only government declarations and notifications but also of flashing news of every kind.

By nature the people are highly inquisitive to enquire about affairs going on around them and express their wanted unwanted commentaries in rapid. For a long time newspapers ruled the realm of information and initiated a curiosity culture, which was fully flourished by the gradual commencement of developed communication techniques. At present multimedia has become the most powerful device of information supply simply we thinks that this affair does not concern to environment, but it should be borne in mind that this is a serious to affect environment as much the physically caused effects do no even. Because the misinterpreted information cause the people to be more combative, and rigours. It is a fact that the egoist thinkers without having any consideration of farsighted results of their expressions become the cause of baseless agitations, as such they sow the seeds of destructive thing to be activated upon. If we go in deep to trace such instances, we can every easily find a lot to ascertain this fact in the history of modern Europe and world. It can be pointed out that ideologies and thoughts have nowadays become highly susceptive to informative prevalence. More and more its fast flow gives it more and more diffusive velocity of infusive nature. For this prupose when information technology, was not in an advanced for thinkers and their followers usually travelled a lot for the pension of their ideologies. This was the original and most trustworthy way of delivering information and spreading ideologies conducive to humanitism and religious followings. It shows that originally expanded culture of man as social being, through involved in struggles and conformations was also too much eager for peace, for which thinkers did a lot, naturally in good faith, as the social thinkers, do in modern times to take human beings out of the uneasing social religious and political condition which prevail upon them due to their self centred attitude. Hence saints like Gautom Buddha and Mahavirji in India and Christ in west appeared, before more than two thousand year.

Writing was the first invention of delivering information. Shang symbols of China are the earliest known examples of writing in the far East, but they were not primitive. They pre suppose a long period of development, Shang Dynasty was represent in 1100 B.C. in China when Cahu Dynasty succeeded Sang Dynasty, it exerted zealous efforts to convince the people that it was legitimate succession rather than a usurpation. This shows the significance of information. If general masses would not have been conscious of informations, they might have revolted needed to make the people believe that the last Shang Ruler had been as incompetent and debauched and that the divine powers had used the Chou as an instrument for his removal. The "Mandate of A Heaven" they alleged had been transferred from the Sang house to the Chou. This proves that since the times immemorial, in general sense too the people realized the impact of informations. Emperor Ashok and other prominent emprors of India were very informative. They caused to setup inscriptions to release informations. However, he festious gatherings, festivals and fairs, periodical markets (haats) etc. were the centres of general informative diffusion. Alongwith this travellers, carvans and pilgrimage were also the conducive participants of information releases. Generally such inormmation medium and means were of high political and religious value. These usually worked as the most influential appliances to affect the masses on one side and cause so many political achievements on the other. The most famous world of Chankya, named "Kautilya Arthashastra" authentically reveals this fact in which he has mentioned how to influence people and were the commercial carvans which caused India famous by the name of 'Bird of Godl'. And thus India was repeatedly invaded by the foreign people. Alluddin, the famous Khalji Sultan of Sultanate Age could know all about populence of Evgiri, from the traders, whose carvans were the usual visitors of that rich kingdom and other of south India. He also enquired of the political conditions and the strength of armed forces before starting to invade that kingdom. Naturally his victory, added a new chapter to Indian History. It happened due to the informations. Alluddin could cause. This caused the heavy influx of Islam in south India. Ultimately to affect Indian culture and its environmental, social and religious set up.

Certain activities of man were also of high informative value in those times, as the construction of idols and temples, shrines, caves, sculpture worked coming there from so many different places. Naturally, these were the centre of fabulous and non-fabulous informations.

Certainly the informations sprang as such, through most of them unauthentic, vague and insignificant but of general living and interest, the christ of public anxiety of getting varied informations, as could be had in the ordinary course of affairs even being alien ultimately causing so many kinds of up heaval effective to enviromental situations. Though, in so many ways man might be unattached to these, but were valuable for gossips and unnecessary arguments with other. For the times immemorial the nature of man has not changed in this respect. Naturally tittle. Tuttle, rumours and fictions even have their roots in its human tendency.

While press and printing technology was not invented people were more fabulous. They usually moved around mythology, religious ideology, historical events and imaginative fables, but printing factor drew people towards the actual sense of life, because publication of newspaper during the period of renaissance in Europe started giving reports of happenings and occurrences, along with the expressions of thinkers regarding the state of affairs of that age. Invention of steam engine, very soon paved the way of replacing the manual postal service and mails became possible to be sent by Railway Mail service. This was the first step of information technology, which gave new fruits to general public. Now ordinary persons could also mail their letters aiming their delivery soon and proper before Railway Main Service. It was quite difficult to have a proper facilitated mail devliery for general public. It can be said that this achievement was most radical in every sense. Hence the society was deeply influenced. The newspapers could be transported to distant places. It means that informations gained a good gap in reaching distant place. Naturally it was the colour beginning of capturing ideas and thinking.

Though such expansion of news service also gave a nearness to people in respect of cultural feelings, but diversities too. Telegraphic system, *i.e.*, telephonic conversation provided much more, and again wireless system and in turn radio and television tool communicative measure very high. These achievements were taken to be very amazing, gradually become very ordinary for masses. They changed the world radically. A revolutionized age emerged, which affects every corner of human life. Agriculture transportation, defence, business, education, medication, industry, banking entertainment art, science etc., all came in the zone of highly sensitive communication technology which have every type of instantly communicative information and knowledge for man at hand. Now you may remain aware of world affairs at every moment. Not only this you can see the live happenings occurring any place in the world on your television screen.

Usually invention of radio, proved to be revolutionary device in the field of transmission of informations. Movies were the another device, which paved the way to invent televisions turn by turn VCD, FAX, internet, appeared as the off shoots of televisions technology. In the vogue of such inventions computer came forward as the sub title life element to dominate the world of knowledge. Now it was also possible to preserve informations in computers and get them relayed through internet, whenever we wish to do so.

In the chain of information technology, its last link is space technology. It is the gift of space service which has instituted so many satellites in space. The signals, which are sent to satellites from the earth are augmented by them, to be returned in man by directions. The augmented singles are captured by the centers established for this purpose on earth and then relayed by them. This is the most developed and amazing achievement of communication technology by this time. The satellites are caused to be stationed at a certain be high in space above the equal to, which relatively move along with the earth remaining in their originally instated position. Space communication system has led the communicative functioning to a great revolution.

Internet. Internet means a net of computers throughout the world. When more than one computer are connected with each other, it so called internet. And when so many such networks are made joint with each other in occupies a vast area to be active within it. Originally, this system of information transmission was instituted by a research agency of American Defence Division. This agency caused four computers connected with each other. Three of these computer were in California, and the further was in Euva. Gradually thirtyseven computer were also jointed with this network. This process was invented for the speedy transmission and receiving of information. This system was lifted up very soon and developed. This caused thereby to its expansion throughout America by 1990. In the year 1996 seven crores of peoples were there connected to internet.

At present there are more than ten lakh internet websites in the world. Every website is in itself an internet, which in trun is connected to some other websites in any way. Internet, actually is the great treasure of informations. These are in their all respects complete, while the informations about any subject cannot be known through newspapers in vividity as these are of temporary nature, and not complete even.

Actually speaking this, internet system possess the world of wisdom and knowledge. Hence there is every reason to denote that it affects the environment in both ways, culturally and physically. The informations which are in the store of internet are open to be borrowed by anybody. But when we talk about the enviormental effects of this advanced technology, these are consequential of the borrowed knowledge. Naturally these can be utilized in a way, injurious to social behaviour, or against ethical and moral standards.

Nationally it may be observed that thoughts, and deliberations communicated to man are not conducive to environmental causes, so far. These do not impel him to prevail upon physical activities. Apparently, this perception seems good, but is not true. This should be born that thoughts are the inciting force, which inclines man to be activated. Thoughts and deliberations of man, though might not be connected to direct physical activities, instantly but these have infusive germs tended to spread, hence possibly the thinkers themselves might not be in direct involvement of getting their ideas fructified, but when these are thrown out, other persons might be following them to be active accordingly. This situation affects the environment. Generally we find that certain persons become leaders due to their deliberations and their observations are followed by others. The same thing can be seen in case of so many writers whose works become the instruments of mass incitement. Maxim Gorky's novel 'Mother' can be cited as good example in this respect, which prevailed upon Russian masses to be persuasive to take active part in Bolshevik Revolution. Usually we can perceive such type of energy in poetic verses, and slogans. Hence it is also a fact that the ideas, generated in the minds are substantial, in respect of the activated of man. The modern technololgy of information is very sharp, possessor of persuasive delicacy, psychologically influencing human tendencies. Above all it is vested with aperceived authenticity. Hence publication of transmission caused by media is supposed to be of high value. Naturally present age can be denoted as 'Media Age'. Because of its being tremendously effective instrument of diverting public opinion. But we can take a note side by side that nowadays media has got its won professional attitude, hence deliberations occurring through it are mysteriously propagative and motivated. We are very sorry to say that due to this reason information technology has more or less tended to be a political weapon, which aims at public opinion, with prejudicial point of view and thereby automatically susceptive to environment.

For example, socialism is the present day talk of the world. With certain diversions, this theory moves around Karl Marx. Generally people do not understand the contemplations of Marx, but they are acquainted with its central idea of equality, which persists them to achieve it through violent means in order to perish capitalists and their mentors. Naturally this theory is most alluring to poor stock, who is the major part of the world population. Hence in concurrence to such contemplations of Marx, European countries were very badly shaken and environment took a sharp turn. Side by side rise of communal nationalism put Europe to a fierce results to be the fructifications of through; diffused by Karl Marx, in sequence of Machiavelli, Thomas Hobbes, John Locke, Rousseau etc. which whirled through Europe. Industrial Revolution and Imperialism was in its background, which provided immense wealth to European countries, but left the poor industrial labour ignored naturally to be influenced by the theory of communism. Ultimately the gradual progress of industrialization in the world caused the labour organized as class, under the industrialization in the world caused the labour organized as class, under the shadow of communistic ideology and environment of world tended to changes. Apprehended by communism thinkers drew out a new course of socialism. Quite near to the communism of Marx, rather can be said to be its offshoot. Hence imperialism, industrialisation, colonization were channelised to flow in new environment of 'Live and Let Live', but colored by selfish motives of attaining political power. Hence socialism was defined in innumerable ways to get the viced concepts of 'Live and Let Live', Liberal by nature man has always remained in search of an everlasting system of social welfare Government, but environment has never allowed him to find it. We can very in rural lowlife. Nobody, can deny now that environment does not have any pratical relation with the ideologies and informations diffused by them order information technology. Actually knowledge diffused through informations, directly affects environment. However, these not always harmful but useful too. Its usefulness can quite apparently and instantly be can be perceived in economic affairs. Information have educative of values, if these can be squeezed properly. Medical science and other science are also in progress nowadays. Information of new researches cannot be ignored form the environmental point of view. Readily received informations are always helpful to administration and of great relief in case of calamities struck people.

Informations are also useful in respect of environmental knowledge. Communication technology, remains in service of human being by releasing every minute environmental situations of the world. It is something very wonderful that even the experts living in any part of this world can serve the mankind as a whole through tele-medicine system. In India Informations Technology Ministry has started work on the development plan of this service. This technology was first taken to service in the year 2001, in Allahabad Mahakumbha Fair.

EXERCISES

1. Write a short note on National Population Policy, 2000.
2. Write main consequences of global warming ?
3. Discuss the issues involved in enforcing the environment laws. How can the awareness be inculcated in public in favour of the green earth ?
4. Write short notes on the following:
   (*a*) Process of climate changing.
   (*b*) Process of weather changing.
   (*c*) Write a short note on utilization of fly ash ?
5. Explain population growth and discuss main causes for it ?
6. How does environment effect the human health ?
7. Write a short note on "Human Rights" ?
8. Write a short note on HIV/AIDS.

❑❑❑

Unit VII FIELD WORK

CURRENT STUDY

An itinary to be made to any nearby mountain, hill, grassland, forest, river. Write a report on the basis of observation and importance. The content of the book (Unit one to seven) how needed information for the study ?

Here the work concurrent is used and study with direct observation in simple cuase practical.

A. Forest

(i) *Background Data:* Note down the name, nature, type (tropical rain forest/bushes etc.) relationship of more biosphere river or sanctuary or national park.

(ii) *Human Impact:* An trophogenic activities, time extraction dam, building deforestation etc.

(iii) *Ecological Utility:* To note down a list of animal section what did they feel colder and humid.

(iv) *Commercial Uses:* Utility of forest on the basis of commercialization.

(v) *Forest Structure:* Salient features, dominant trees, density growth or degraded.

B. Mountatin/Hilly

(i) *Background data:* Note down the name of mountain/hill area, average annual rainfall, temperature.

(ii) *Observation of Natural Agitations:* Make a note on forest plashes, density, deforestation, dominant trees, sheds etc. Names of the trees and utility of local route.

(iii) *Plantation:* Type of plantation, type of farming which is going on it may please shifting, traditional cultivation etc., terrace farming, strip contour cropping and its importance.

Nutrient wasteness of these corps will routed to hill environment, human activities, tourism, mining, quarrying, construction on estate.

(*iv*) *Grassland:* Type of grassland, annual periodical tall or short grasses, protected or financed identification of dominant plants or grasses, the dominant plants have course, hard texture earth shaped, thorn type of roots, runner fibres, adipose, write a note on soil erosion, grazing or overgrazing;

(*v*) *Utility.* Make a list of the utilities of the grassland.

RIVER ENVIRONMENT

Background Data: Name the river or tributary and its origin place, to know the river is seasonal or perennial.

(*a*) Water Quality Observation

Purity of water expectation of society regarding purity of water or air. Note down the probable impacts of thermal pollution of aquatic life. To find out sources of pollutions in river pH value of water is required.

(*b*) Observations on Aquatic life—make a food diagram of river or aquatic animals.

Aquatic Animals

Uses: make a list to use water by aquatic animals.

(c) Polluted Site

Human activities is directly related to each pollutions of the environment due to the urbanization and industrialization. Agricultural places make pesticides pollution, waterlogging. Any mysteries area will be very important to know about polluted site.

Industrially Polluted Area

(*i*) *Background Data:* Name of the industry, its capacity, year of establishment and their product.

(*ii*) *Pollution Effect:* Details of stack chimney, toxic gases are produced by the same observation of huge helps of sludge around outside the industries, ability quality and maintence of effluent treatment plant (ETP) in the industry.

(*iii*) *Green Belt:* Green belt in industry is mandatory for industry importance.

(*iv*) *Health Effect.* Health impacts in the human being living with vicinity of the industry.

Common Plants, Insects and Birds

(*i*) *Plants:* To find out and to study the common plants of your area including herbs, trees etc.

(*ii*) *Medicinal Plants:* Local human being often have knowledge about plants for medicinal values, make a list of plants for the same.

(*iii*) *Timberwood Trees:* To note down the important list of tree which yield timber wood.

(*iv*) *Insects:* To identify some common insects of your community.

(*v*) *Birds:* Make a list of useful and un-useful birds of your locality.

A Waterlogged/saline Land

(*i*) *Background Information:* To visit waterlogged salt affected land in rural area farmer historical background, is it agricultural land converted in waterlogged and saline ?

(*ii*) *Remediation:* To find out what remedial measures is to be taken by farmers for the same as well as measures can be suggested.

□□□

APPENDIX

WATERBORNE DISEASE CAUSING ORGANISM

(A) Bacterial

S.No.	Organism Name	Disease	Main Source
1.	*Vibrio cholerae*	Cholera	Human faeces
2.	*Yersinia enterocolitica*	Gastroenteritis	Human and animal faeces
3.	*Salmonella typhi*	Typhoid fever	Human faeces
4.	*Legionella pneumophila* and related bacteria	Acute respiratory illness	Thermally enriched water
5.	Opportunistic	Variable	Natural water
6.	*Shigella*	Bacillary dysentery	Human faeces
7.	*Campylobacter tejuni*	Gastroenteritis	Human and animal faeces
8.	*Mycobacterium Tuberculosis*	Tuberulosis	Human respiratory exudates
9.	Enteropathogenic *E. Coli*	Gastroenteritis	Human faeces
10.	*Salmonella Paratyphi*	Paratyphoid fever	Human faeces

(B) Viruses

S.No.	Organism Name	Disease	Main Source
1.	Rota viruses	Gastroenteritis	Human faeces
2.	Hepatitis A Virus	Infectious hepatitis	Human faeces
3.	Enteroviruses Poliviruses	Poliomyelitis	Human faeces
4.	Echo Viruses	Aseptic meningitis	Human faeces
5.	Coxsackie Viruses A	Aseptic meningitis	Human faeces
6.	Coxsackie Viruses B	Aseptic meningitis	Human faeces
7.	Norwalk and related gastrointestinal viruses	Gastroenteritis	Human faeces
8.	Adeno viruses	Upper respiratory and gastrointestinal illness	Human faeces

(C) Algae

1.	*Schizothrix calciola*	Gastroenteritis	Natural water
2.	*Anabaena flosaquae*	Gastroenteritis	Natural water
3.	*Microsystic acruginosa*	Gastroenteritis	Natural water
4.	*Aphanizomenon flosaquae*	Gastroenteritis	Natural water

(D) Protozoa

1.	*Giardia lamblia*	Giardlasis	Human and animal faeces
2.	*Cryptosporidium*	Cryptosporidiosis	Human and animal faeces
3.	*Entamoeba histolytica*	Amoebic dysentery	Human faeces
4.	*Balantidium cali*	Dysentery	Human faeces

GLOSSARY

[A]

Abatement	:	Reducting pollution through reuse or waste treatment.
Abiotic	:	Non-living.
Abiotic component	:	Non-living part of the environment.
Acid mine drainage	:	Mineral acids that may leach from mining operations to pollute ground.
Act	:	A government rules.
Adsorption	:	Attachment of gas molecules in solution to the surface of solid materials with which they come in contact.
Aerobic	:	An organism that needs oxygen to carry on.
Aerobid	:	Characterized by the presence of free oxygen.
Aesthetic	:	Pleasing to the eye.
AIDS	:	Acquired Immuno Deficiency Syndrome.
Allergens	:	Substance causing allergy.
Alpha biodiversity	:	Biodiversity within community.
Altitude	:	Height (above sea level).
Ambient air	:	The air surrounding.
Annual	:	A year.
Arid	:	Dry area.
Ash	:	Mineral content of a product that remains after complete combustion.

[B]

Bacteria	:	Group of prokaryotic organism except blue green algae.
Benthos	:	Organisms living on.
Biotic	:	Living.

[C]

Carnivore	:	Organism that feed on the other animals.
Cell	:	The structural and functional unit of most living organism.

CTBT	:	Comprehensive Test Ban Treaty in 1996.
Core	:	Inner zone of the Earth.
Crust	:	Solid and the outer layer of the Earth.
Cyanobacteria	:	Blue green algae.

[D]

DNA	:	Deoxyribonucleic acid.
Desert	:	A biome where evaporation exceeds precipitation.
Detritus	:	Dead organic matter.
DDT	:	Dichloro Diphenyl Trichloroethane.

[E]

Effluent	:	A common term for a fluid emitted from a source.
Energy	:	The capacity to perform work.
ELISA test	:	Enzyme linked Immuno Sorbent Test (AIDS Testing).
Ethics	:	Moral values and principles to guidens.
Exotic species	:	New species entering into a geographical area.

[F]

Famine	:	Acute food shortage.
Funa	:	All the animals present in a given area.
Food web	:	A system of food chains that are linked with one another.
Forest	:	A biome in which dominant plants are trees.
Flora	:	All the plants present in a given area.
Fly ash	:	Very fine particles resulting from the burning of fuels like as coal.
Fungicides	:	Chemicals that kill fungi.

[G]

Gene	:	A unit of heredity, it is either DNA or RNA.
Game species	:	Wild animals hunted.
Gasohol	:	A fuel that is a mixture of gasoline and alcohol.
Gamma (rays)	:	Very short wavelength ionizing rays with a high energy.
Geothermal energy	:	Energy derived from the internal heat of the Earth.
Glacier	:	A flowing body of ice.
GDP	:	Gross Domestic Product.
GNP	:	Gross National Product.
Growth rate	:	The increase in population.
Gully erosion	:	Removal of layers of soil creating large channels.

[H]

Habitat : Place where an organism lives.

Herbivore : Plant eater.

Half time : The time required for half the original amount of a substance to disappear.

Herbicide : A chemical used to control unwanted herbaceous plants.

HIV : Human Immuno Deficiency Virus.

Hot spot : Species rich most threatened reservoir of plants and animals on the Earth.

Hydrophyte : A plant living in water.

Hydrosere : A sere beginning in a wet habitat.

[I]

Infiltration : Percolation of water into the soil.

Invertebrates : Animals that have no back bone.

Ion : Atoms with positive charge.

Insolation : Incoming solar radiation.

[J]

Joule : A unit for energy (4.184 Joules = 1 Calories).

Jhum cultivation : Shifting cultivation.

[L]

Landslides : Down hills.

Latitude : Distance from the equator.

Lagoon : A large shallow pond.

Litter : Waste matter deposited on land.

Lentic water : Standing water.

Lithosphere : The crust and mantle of the Earth.

[M]

Magma : Molten rock below the Earth.

MBA : Man and biosphere.

Marsh : A wet land without trees.

Manure : The excreta of animals.

Macrophytes : Large aquatic plants.

Malnutrition : Diet with deficiency of proteins.

Matter : Anything that has mass.

Mutation : A sudden heritable change.

MIC : Methyl isocyanate.

[N]

NEERI	:	National Environmental Engineering Research Institute, Nagpur.
Non-conventional Energy	:	New sources of energy.

[P]

Population	:	Group of individual organism of a species living within a particular region.
Population explosion	:	Exponential growth of population to size that exceeds the carrying capacity.
Primary pollutants	:	Chemicals released directly into the air.
Primary succession	:	Succession that begins on a sterile region where conditions of existence are not at first favourable.
Protozoans	:	A group of unicellular or acellular.
Psammosere	:	Complete set of communities that develop in sandy habitat during succession and ultimately stabilise such habitat.

[R]

Radiant energy	:	Solar energy travelling in the form of radiations.
Red data book	:	A compilation of data on species threatened with extinction, maintained by IUCN.
Remediation	:	Cleaning up chemical contaminants from polluted area.
Renewable energy	:	Sources which are generated by the natural processes, so that they can be used indefinitely.
Run off	:	Water from rain or snow that comes down the land surface into streams and rivers.

[S]

Sacred forests	:	Forest patches protected by tribal communities due to religious sanctity accorded to them.
Salinity	:	Amount of soluble salts in water or soil.
Sand	:	Soil particles between 50–200 microns in diameter.
Scrubber	:	An equipment used to clean air for both dusts and gases by passing it through a dry or wet packing material.
Secondary pollutant	:	A pollutant formed by the combination of other pollutants in the environment.
Secondary productivity	:	Rate of energy storage at consumer's level.

Secondary consumer	:	The community development on sites which are previously occupied by well developed communities.
Sedimentary cycle	:	Nutrient cycle's reservoir exists in the earth's crust.
Soil	:	Weathered surface material.
Solar system	:	The sun, the Planets and the Satellites.
Soil erosion	:	The Loss of fertile top soil.
Species	:	Fundamental unit of study in taxonomy, denoting particular interbreeding group of organisms.
Species diversity	:	It refers to the variety of species within a region.
Standing quality	:	The number of inorganic substances present at any given time in the environment of an ecosystem.
Stress	:	Such factors that causes a strain on an organism.
Stratosphere	:	The region of atmosphere above the troposphere in which temperature rises with height.
Succession	:	A natural process by which different communities colonise the same area over a period of time in definite sequence.

[T]

Toilings	:	Mining wastes.
Thermodynamics	:	A branch that deals with transfer and conversions of energy.
Toxins	:	Poisonous chemicals, harmful even in small concentrations.
Troposphere	:	The lowest level of atmosphere in which temperature decreases with height. This is about 20 km above earth's surface.

[U]

Upwelling	:	Movement of nutrient rich bottom water to sea's surface.
UNESCO	:	United Nations Educational, Scientific and Cultural Organisation.
UNEP	:	United Nations Environment Programme.
Urbanization	:	Increasing concentration of population in cities.

[V]

Vertebrates	:	Animals with back bones.

[W]

Water logging	:	Excessive saturation of soil with water making it difficult for plant roots to take hold.

Water pollution	:	Addition to any inorganic and organic substances, which may affect the physio-chemical and biological properties of water, rendering to unfit for a particular use.
Watershed	:	The land area from which water drains under gravity to a common drainage channel.
Wildlife	:	Undomesticated life forms.
WWF	:	World Wide Fund.

[X]

Xerophyte	:	A plant adopted to grow under dry conditions.
X-ray	:	Very short wavelength rays, useful in medical diagnosis can cause mutation.

[Z]

Zooplankton	:	Microscopic floating animals in surface of water feeding on the phytoplarktons in aquatic ecosystem.

❑❑❑

EXERCISES

UNIT I

1. What is the meaning of Environment ? Discuss its scope and importance in short ?
2. How environmental studies is beneficial to us ?
3. Write short notes on:
   (*a*) Hydrosphere (*b*) Atmosphere
   (*c*) Lithosphere (*d*) Biosphere

UNIT II

1. What is meant by Natural Resources ? Write its main types ?
2. Give a detailed account of Forest Resources ?
3. Write a detailed note on Water Resources ?
4. Write a detailed note on Mineral Resources ?
5. Write short note on Marine Mineral Resources ?
6. Write a detailed note on Food Resources ?
7. Write a detailed note on Energy Resources ?
8. Discuss the Land Resources ?

UNIT III

1. What is the meaning of "Ecosystem" ? Define it.
2. Write different types of ecosystem.
3. Write detailed note on "Ecological Succession".
4. Write short note on Food Chain.
5. Discuss a short note on Food Web.
6. Write a detailed note on "Ecological Pyramids".
7. Write short notes on:
   (*a*) Grassland Ecosystem (*b*) Forest Ecosystem
   (*c*) Desert Ecosystem (*d*) Aquatic Ecosystem
   (*e*) Ocean Ecosystem.

UNIT IV

1. Write defined note on biodiversity and its conservation.
2. Write a note on "biogeographical classification of India".
3. Write short notes on the following:
   (*a*) Biodiversity at global level
   (*b*) Biodiversity at national level
   (*c*) India as a mega-diversity nation
   (*d*) Hot spots of biodiversity.
4. Write a short note on endemic species of India.

UNIT V

1. What is the meaning of pollution ? Write main types of pollution.
2. What is the meaning of air pollution ? Write a short note on air pollution.
3. What is the meaning of soil pollution ? Write its main causes and effects.
4. What is the meaning of ocean pollution ? Write its main causes and effects.
5. What is the meaning of noise pollution ? Write main causes of noise pollution.
6. What is the meaning of water pollution ? Write important causes and effects of water pollution.
7. What is the meaning of thermal pollution ? Write its causes and effects ?

UNIT VI

1. Explain recent approaches to sustainable development.
2. Discuss the urban problems related to energy.
3. Write a short note on water conservation.
4. Explain rainwater harvesting.
5. Discuss the problems and concerns of resettlement and rehabilitation of people ?
6. What are issues and possible situations related to environment ethics ?
7. Write short notes on these:
   (*a*) Climate change (*b*) Global warming
   (*c*) Ozone layer (*d*) Acid rain.
8. Write short notes on these:
   (*a*) The Air Act, 1981 (*b*) Water Act, 1974
   (*c*) Wildlife Act, 1972 (*d*) Forest Act, 1980.
9. Write a short note on ecological forming system ?

2. Write a short note on National Population Policy, 2000.
2. Write main consequences of global warming ?
3. Discuss the issues involved in enforcing the environment laws. How can the awareness be inculcated in public in favour of the green earth ?
4. Write short notes on the following:
   (*a*) Process of climate changing.
   (*b*) Process of weather changing.
   (*c*) Write a short note on utilization of fly ash ?
5. Explain population growth and discuss main causes for it ?
6. How does environment effect the human health ?
7. Write a short note on "Human Rights" ?
8. Write a short note on HIV/AIDS.

❑❑❑